Pitman Research Notes in Mathematics Series

Submission of proposals for consideration
Suggestions for publication, in the form of outlines and representative samples, are invited by the Editorial Board for assessment. Intending authors should approach one of the main editors or another member of the Editorial Board, citing the relevant AMS subject classifications. Alternatively, outlines may be sent directly to the publisher's offices. Refereeing is by members of the board and other mathematical authorities in the topic concerned, throughout the world.

Preparation of accepted manuscripts
On acceptance of a proposal, the publisher will supply full instructions for the preparation of manuscripts in a form suitable for direct photo-lithographic reproduction. Specially printed grid sheets are provided and a contribution is offered by the publisher towards the cost of typing. Word processor output, subject to the publisher's approval, is also acceptable.

Illustrations should be prepared by the authors, ready for direct reproduction without further improvement. The use of hand-drawn symbols should be avoided wherever possible, in order to maintain maximum clarity of the text.

The publisher will be pleased to give any guidance necessary during the preparation of a typescript, and will be happy to answer any queries.

Important note
In order to avoid later retyping, intending authors are strongly urged not to begin final preparation of a typescript before receiving the publisher's guidelines and special paper. In this way it is hoped to preserve the uniform appearance of the series.

Longman Scientific & Technical
Longman House
Burnt Mill
Harlow, Essex, UK
(tel (0279) 26721)

Longman Scientific & Technical
Churchill Livingstone Inc.
1560 Broadway
New York, NY 10036, USA
(tel (212) 819-5453)

Titles in this series

1 Improperly posed boundary value problems
A Carasso and A P Stone
2 Lie algebras generated by finite dimensional ideals
I N Stewart
3 Bifurcation problems in nonlinear elasticity
R W Dickey
4 Partial differential equations in the complex domain
D L Colton
5 Quasilinear hyperbolic systems and waves
A Jeffrey
6 Solution of boundary value problems by the method of integral operators
D L Colton
7 Taylor expansions and catastrophes
T Poston and I N Stewart
8 Function theoretic methods in differential equations
R P Gilbert and R J Weinacht
9 Differential topology with a view to applications
D R J Chillingworth
10 Characteristic classes of foliations
H V Pittie
11 Stochastic integration and generalized martingales
A U Kussmaul
12 Zeta-functions: An introduction to algebraic geometry
A D Thomas
13 Explicit *a priori* inequalities with applications to boundary value problems
V G Sigillito
14 Nonlinear diffusion
W E Fitzgibbon III and H F Walker
15 Unsolved problems concerning lattice points
J Hammer
16 Edge-colourings of graphs
S Fiorini and R J Wilson
17 Nonlinear analysis and mechanics: Heriot-Watt Symposium Volume I
R J Knops
18 Actions of fine abelian groups
C Kosniowski
19 Closed graph theorems and webbed spaces
M De Wilde
20 Singular perturbation techniques applied to integro-differential equations
H Grabmüller
21 Retarded functional differential equations: A global point of view
S E A Mohammed
22 Multiparameter spectral theory in Hilbert space
B D Sleeman
24 Mathematical modelling techniques
R Aris
25 Singular points of smooth mappings
C G Gibson
26 Nonlinear evolution equations solvable by the spectral transform
F Calogero
27 Nonlinear analysis and mechanics: Heriot-Watt Symposium Volume II
R J Knops
28 Constructive functional analysis
D S Bridges
29 Elongational flows: Aspects of the behaviour of model elasticoviscous fluids
C J S Petrie
30 Nonlinear analysis and mechanics: Heriot-Watt Symposium Volume III
R J Knops
31 Fractional calculus and integral transforms of generalized functions
A C McBride
32 Complex manifold techniques in theoretical physics
D E Lerner and P D Sommers
33 Hilbert's third problem: scissors congruence
C-H Sah
34 Graph theory and combinatorics
R J Wilson
35 The Tricomi equation with applications to the theory of plane transonic flow
A R Manwell
36 Abstract differential equations
S D Zaidman
37 Advances in twistor theory
L P Hughston and R S Ward
38 Operator theory and functional analysis
I Erdelyi
39 Nonlinear analysis and mechanics: Heriot-Watt Symposium Volume IV
R J Knops
40 Singular systems of differential equations
S L Campbell
41 N-dimensional crystallography
R L E Schwarzenberger
42 Nonlinear partial differential equations in physical problems
D Graffi
43 Shifts and periodicity for right invertible operators
D Przeworska-Rolewicz
44 Rings with chain conditions
A W Chatters and C R Hajarnavis
45 Moduli, deformations and classifications of compact complex manifolds
D Sundararaman
46 Nonlinear problems of analysis in geometry and mechanics
M Atteia, D Bancel and I Gumowski
47 Algorithmic methods in optimal control
W A Gruver and E Sachs
48 Abstract Cauchy problems and functional differential equations
F Kappel and W Schappacher
49 Sequence spaces
W H Ruckle
50 Recent contributions to nonlinear partial differential equations
H Berestycki and H Brezis
51 Subnormal operators
J B Conway

52 Wave propagation in viscoelastic media
F Mainardi
53 Nonlinear partial differential equations and their applications: Collège de France Seminar. Volume I
H Brezis and J L Lions
54 Geometry of Coxeter groups
H Hiller
55 Cusps of Gauss mappings
T Banchoff, T Gaffney and C McCrory
56 An approach to algebraic K-theory
A J Berrick
57 Convex analysis and optimization
J-P Aubin and R B Vintner
58 Convex analysis with applications in the differentiation of convex functions
J R Giles
59 Weak and variational methods for moving boundary problems
C M Elliott and J R Ockendon
60 Nonlinear partial differential equations and their applications: Collège de France Seminar. Volume II
H Brezis and J L Lions
61 Singular systems of differential equations II
S L Campbell
62 Rates of convergence in the central limit theorem
Peter Hall
63 Solution of differential equations by means of one-parameter groups
J M Hill
64 Hankel operators on Hilbert space
S C Power
65 Schrödinger-type operators with continuous spectra
M S P Eastham and H Kalf
66 Recent applications of generalized inverses
S L Campbell
67 Riesz and Fredholm theory in Banach algebra
B A Barnes, G J Murphy, M R F Smyth and T T West
68 Evolution equations and their applications
F Kappel and W Schappacher
69 Generalized solutions of Hamilton-Jacobi equations
P L Lions
70 Nonlinear partial differential equations and their applications: Collège de France Seminar. Volume III
H Brezis and J L Lions
71 Spectral theory and wave operators for the Schrödinger equation
A M Berthier
72 Approximation of Hilbert space operators I
D A Herrero
73 Vector valued Nevanlinna Theory
H J W Ziegler
74 Instability, nonexistence and weighted energy methods in fluid dynamics and related theories
B Straughan
75 Local bifurcation and symmetry
A Vanderbauwhede
76 Clifford analysis
F Brackx, R Delanghe and F Sommen
77 Nonlinear equivalence, reduction of PDEs to ODEs and fast convergent numerical methods
E E Rosinger
78 Free boundary problems, theory and applications. Volume I
A Fasano and M Primicerio
79 Free boundary problems, theory and applications. Volume II
A Fasano and M Primicerio
80 Symplectic geometry
A Crumeyrolle and J Grifone
81 An algorithmic analysis of a communication model with retransmission of flawed messages
D M Lucantoni
82 Geometric games and their applications
W H Ruckle
83 Additive groups of rings
S Feigelstock
84 Nonlinear partial differential equations and their applications: Collège de France Seminar. Volume IV
H Brezis and J L Lions
85 Multiplicative functionals on topological algebras
T Husain
86 Hamilton-Jacobi equations in Hilbert spaces
V Barbu and G Da Prato
87 Harmonic maps with symmetry, harmonic morphisms and deformations of metrics
P Baird
88 Similarity solutions of nonlinear partial differential equations
L Dresner
89 Contributions to nonlinear partial differential equations
C Bardos, A Damlamian, J I Díaz and J Hernández
90 Banach and Hilbert spaces of vector-valued functions
J Burbea and P Masani
91 Control and observation of neutral systems
D Salamon
92 Banach bundles, Banach modules and automorphisms of C*-algebras
M J Dupré and R M Gillette
93 Nonlinear partial differential equations and their applications: Collège de France Seminar. Volume V
H Brezis and J L Lions
94 Computer algebra in applied mathematics: an introduction to MACSYMA
R H Rand
95 Advances in nonlinear waves. Volume I
L Debnath
96 FC-groups
M J Tomkinson
97 Topics in relaxation and ellipsoidal methods
M Akgül
98 Analogue of the group algebra for topological semigroups
H Dzinotyiweyi
99 Stochastic functional differential equations
S E A Mohammed

151 A stochastic maximum principle for optimal control of diffusions
U G Haussmann
152 Semigroups, theory and applications. Volume II
H Brezis, M G Crandall and F Kappel
153 A general theory of integration in function spaces
P Muldowney
154 Oakland Conference on partial differential equations and applied mathematics
L R Bragg and J W Dettman
155 Contributions to nonlinear partial differential equations. Volume II
J I Díaz and P L Lions
156 Semigroups of linear operators: an introduction
A C McBride
157 Ordinary and partial differential equations
B D Sleeman and R J Jarvis
158 Hyperbolic equations
F Colombini and M K V Murthy
159 Linear topologies on a ring: an overview
J S Golan
160 Dynamical systems and bifurcation theory
M I Camacho, M J Pacifico and F Takens
161 Branched coverings and algebraic functions
M Namba
162 Perturbation bounds for matrix eigenvalues
R Bhatia
163 Defect minimization in operator equations: theory and applications
R Reemtsen
164 Multidimensional Brownian excursions and potential theory
K Burdzy
165 Viscosity solutions and optimal control
R J Elliott
166 Nonlinear partial differential equations and their applications. Collège de France Seminar. Volume VIII
H Brezis and J L Lions
167 Theory and applications of inverse problems
H Haario
168 Energy stability and convection
G P Galdi and B Straughan
169 Additive groups of rings. Volume II
S Feigelstock
170 Numerical analysis 1987
D F Griffiths and G A Watson
171 Surveys of some recent results in operator theory. Volume I
J B Conway and B B Morrel
172 Amenable Banach algebras
J-P Pier
173 Pseudo-orbits of contact forms
A Bahri
174 Poisson algebras and Poisson manifolds
K H Bhaskara and K Viswanath
175 Maximum principles and eigenvalue problems in partial differential equations
P W Schaefer
176 Mathematical analysis of nonlinear, dynamic processes
K U Grusa
177 Cordes' two-parameter spectral representation theory
D F McGhee and R H Picard
178 Equivariant K-theory for proper actions
N C Phillips
179 Elliptic operators, topology and asymptotic methods
J Roe

D F McGhee & R H Picard

University of Strathclyde / University of Wisconsin–Milwaukee

Cordes' two-parameter spectral representation theory

Longman Scientific & Technical

Copublished in the United States with
John Wiley & Sons, Inc., New York

Longman Scientific & Technical
Longman Group UK Limited
Longman House, Burnt Mill, Harlow
Essex CM20 2JE, England
and Associated Companies throughout the world.

Copublished in the United States with
John Wiley & Sons, Inc., 605 Third Avenue, New York, NY 10158

First published 1988

AMS Subject Classifications: (main) 47A50, 47B25
(subsidiary) 34B25, 46C10, 81C10

ISSN 0269-3674

British Library Cataloguing in Publication Data
McGhee, D. F.
Cordes' two-parameter spectral representation theory.
1. Hilbert spaces. Self-adjoint operators.
Spectoral representation
I. Title II. Picard, R.
515.7'33
ISBN 0-582-02342-4

Library of Congress Cataloging-in-Publication Data
McGhee, D. F.
Cordes' two-parameter spectral representation theory.
(Pitman research notes in mathematics series,
ISSN 0269-3674 ; 177)
Bibliography: p.
1. Hilbert space. 2. Linear operators. 3. Spectral
theory (Mathematics) I. Picard, R. II. Title.
III. Title: 2-parameter spectral representation theory.
IV. Series.
QA322.4.M43 1988 515.7'33 87-36683

ISBN 0-470-21084-2 (USA only)

Printed and bound in Great Britain by
Biddles Ltd, Guildford and King's Lynn

Contents

Preface

These notes have been developed from a series of research lectures given by the authors at various times at the University of Calgary and the University of Bonn. They deal with a class of two-parameter problems in Hilbert space that was first considered by H.O. Cordes in the early 1950's. These problems can be classed as part of what is now referred to as multiparameter spectral theory.

One of the main aims of multiparameter spectral theory is to generalise to this setting the major results concerning the spectrum of a self-adjoint operator in Hilbert space. In particular the multiparameter generalisation of the spectral representation theorem for a self-adjoint operator is of great interest.

General results in this area have not been readily achieved. For a so-called strongly right-definite system involving bounded self-adjoint operators a result has been well known for some time (see [9, Chapter 3]). For such a system involving unbounded operators, a similar result has been achieved (see [9, Chapter 4] and [10]) but its development proved somewhat more troublesome.

However, these results are not entirely satisfactory. The spectral representations are achieved, not in terms of the properties of the operators in the multiparameter problem as it is originally posed, but instead in terms of a somewhat complicated set of commuting operators which are constructed from the original system. This in a sense amounts to setting out to solve a partial differential equation by a separation of variable technique, and attempting to make progress by reversing this process to reconstitute the original equation.

In the 1950's H.O. Cordes published a series of lengthy research papers on the method of separation of variables studied in abstract Hilbert space [5,6]. Within these works, he achieved, for a two-parameter problem, a spectral representation in terms of the properties of the original operators in the system. To our knowledge, Cordes' results do not follow in their entirety from any of the results concerning multiparameter spectral representations which have been achieved more recently. In particular, the work of Berezanskii [4] deals with a special case where one of the two operators depends on only one of the parameters.

It is the intention of these notes to bring to a wide audience the results concerning two-parameter spectral representations which were achieved by Cordes. We present the work in terminology and notation that is familiar to present day workers in operator theory in general and multiparameter spectral theory in particular.

It is assumed that the reader has a good background in functional analysis and the theory of self-adjoint operators in Hilbert space. The spectral theorem for a self-adjoint operator in Hilbert space is assumed throughout.

The authors are deeply indebted to the University of Calgary and, in particular, to Professor P.J. Browne and Professor P.A. Binding for financial assistance for study visits and for providing an inspiring environment and encouragement for the development of these notes. Acknowledgement is due to the Sonderforschungsbereich 72, University of Bonn, for identical reasons.

Also, we would like to express our sincere thanks to Professor Rolf Leis, University of Bonn, and to Professor Gary Roach, University of Strathclyde, for being inspirational colleagues and friends to both of us.

Finally, of course, we acknowledge gratefully the expertise of Miss Mary Doherty who prepared the complete typescript.

Notation

In general, symbols are defined when used in the text. However, the following are used without explanation:

$\mathcal{D}(\cdot)$	the domain of a linear operator
Id	the identity operator, often supressed when multiplied by a constant, e.g. $T-\lambda \equiv T-\lambda\mathrm{Id}$.
$N(\cdot)$	the null space of a linear operator
$R(\cdot)$	the range of a linear operator
$\rho(\cdot)$	the resolvent set of a linear operator
$\sigma(\cdot)$	the spectrum of a linear operator
$\sigma_p(\cdot)$	the point spectrum of a linear operator
$\mathrm{Sp}[\cdot]$	the linear span of a set of elements
$T\|_M$	the restriction of an operator T to a subspace $M \subset \mathcal{D}(T)$
$\subset\subset$	compact subset
$\perp(\perp_H)$	orthogonal complement (with, if required for clarity, a note of the space in which it is being taken).

1 Multiparameter spectral theory: an introduction

Spectral theory for a linear operator T defined in a Hilbert space H is the study of properties of the operator

$$T - \lambda \mathrm{Id} : H \supset \mathcal{D}(T) \to H \tag{1.1}$$

where Id is the identity operator on H and $\lambda \in \mathbb{C}$ is the spectral parameter. The generalisation of this situation which has become known as multiparameter spectral theory concerns systems of linear operators of the form

$$W_r(\underset{\sim}{\lambda}) := T_r - \sum_{r=1}^{k} \lambda_s V_{rs} : H_r \supset \mathcal{D}(W_r(\underset{\sim}{\lambda})) \to H_r, \quad r = 1,\ldots,k, \tag{1.2}$$

where T_r, V_{rs}, $s = 1,\ldots,k$ are linear operators defined in Hilbert spaces H_r, $r = 1,\ldots,k$, and $\underset{\sim}{\lambda} := (\lambda_1,\ldots,\lambda_k) \in \mathbb{C}^k$ is a k-tuple of spectral parameters linking the k operators in (1.2). The aim of the theory is to extend to this multiparameter setting the well known and extremely important result concerning the spectrum of the standard one-parameter operator (1.1).

Much of the motivation for the recent work on multiparameter spectral theory came from an address by F.V. Atkinson delivered to a meeting of the American Mathematical Society in 1965 and subsequently published in its Bulletin in 1968 [1]. This was followed by the same author's definitive work on multiparameter problems involving matrices and compact operators [2]. However, there had been a great deal of earlier work on such problems and we cite [1,2] for a comprehensive list of references.

Of particular interest to us here is the multiparameter generalisation of the spectral representation theorem for a self-adjoint operator in Hilbert space. Much of the more recent work on multiparameter theory has been devoted to this area. Here, we consider systems of operators of the form (1.2) with $T_r: H_r \supset \mathcal{D}(T_r) \to H_r$ self-adjoint and $V_{rs}: H_r \to H_r$ bounded self-adjoint, $s = 1,\ldots,k$, $r = 1,\ldots,k$.

Various results in this direction have been achieved and can be found in the monograph of Sleeman [9]. However, all of these give a spectral representation for the system (1.2) in terms of the spectral families of a somewhat complicated set of commuting operators defined in terms of T_r, V_{rs}, $s = 1,\ldots,k$, $r = 1,\ldots,k$, and acting in a weighted version of the tensor product space

$$H^{\otimes} := \bigotimes_{r=1}^{k} H_r .$$

However, since one of the important motivations for the study of multiparameter problems is the method of separation of variables for partial differential equations (where the spectral parameters appear as constants of separation), and the tensor product construction is in fact the reverse of this process, then it is clear that a spectral representation in terms of the spectral families of these commuting operators in $H^{\otimes}$ is not what one would expect to be useful in practice. Instead we would hope for a spectral representation in terms of the 'separated' and hence simpler operators appearing in (1.2).

Results in this direction for a two-parameter problem, due to H.O. Cordes [5,6] have been known for some time. It is this work which is the subject of these notes. Little attempt has been made to develop Cordes' approach, perhaps because of the extremely complicated nature of the calculations which appear to be necessary. However, to our knowledge the results achieved cannot be deduced from any of the more recent multiparameter spectral representation theorems.

We attempt to develop the theory in a notation and style that will be familiar to present day workers in the field. Section 2 states and proves various lemmas concerning linear operators and their inverses on weighted versions of a Hilbert space. Section 3 introduces the two-parameter system to be discussed and gives the usual construction of the associated pair of operators in the tensor product space (denoted as usual by Γ_1 and Γ_2).

Sections 5 and 6 show that these operators are essentially self-adjoint and commute and Section 7 develops the two-parameter spectral theorem in terms of the spectral families associated with the closures of these operators. This is the position usually achieved in more recent developments for multiparameter systems.

Sections 8 to 10 shows how this spectral representation can be expressed in terms of the spectral families of the original self-adjoint operators given the further assumption that one of the self-adjoint operators T_1 or T_2 has a purely discrete spectrum.

2 Preliminary results

In this section we give various lemmas that are required for later analysis.

Let H be a separable Hilbert space with inner product $(.,.)$ and norm $\|\cdot\| := \sqrt{(.,.)}$.

Definition 2.1: A linear operator $V : H \supset \mathcal{D}(V) \to H$ is said to be

(i) positive definite, denoted by $V > 0$, if $(Vf,f) > 0$, $\forall 0 \neq f \in \mathcal{D}(V)$;

(ii) strongly positive definite, denoted by $V >> 0$, if $\exists c > 0$ such that $(Vf,f) \geq c\|f\|^2$ $\forall f \in \mathcal{D}(V)$.

Lemma 2.2: Let $V : H \to H$ be a bounded positive definite linear operator on H. Then

(i) $[.,.] := (.,V.)$ defines an inner product on H;

(ii) if $\mathcal{H}$ denotes the completion of H with respect to the norm $|||\cdot||| := \sqrt{[.,.]}$, and if $W : H \to H$ is a bounded linear operator that commutes with V, then W can be extended to all of $\mathcal{H}$ as a bounded linear operator and $|||W||| \leq \|W\|$. Further, if W is positive definite on H then its extension is positive definite on $\mathcal{H}$;

(iii) $V\mathcal{H} \subset H$ so that $\mathcal{D}(V^{-1}) \subset H$.

Proof: (i) This follows from the positive definiteness of V.

(ii) Since $W : H \to H$ commutes with V, it follows that $\forall f \in H$

$$\begin{aligned}
|||Wf|||^2 &= (Wf, VWf) \\
&= (V^{1/2}Wf, V^{1/2}Wf) \\
&= (WV^{1/2}f, WV^{1/2}f) \\
&\leq \|W\|^2 (f, Vf) \\
&= \|W\|^2 \; |||f|||^2,
\end{aligned}$$

so that $W : \mathcal{H} \supset H \to \mathcal{H}$ is bounded with $|||W||| \leq \|W\|$. Since H is dense in $\mathcal{H}$, W can be extended uniquely by continuity to all of $\mathcal{H}$ and this extended operator, which we will continue to denote by W, has the same norm. Now, if W is positive definite on H it is a simple calculation to show that W is self-adjoint on $\mathcal{H}$. Since W is self-adjoint on H and $0 \notin \sigma_p(W)$, the point spectrum of W, WH is dense in H. It is easy to show that a set that is dense in H is also dense in $\mathcal{H}$, so that WH is dense in $\mathcal{H}$. Now, suppose that $Wf = 0 \in \mathcal{H}$ for some $f \in \mathcal{H}$. Then, $\forall g \in H$

$$0 = [g, Wf] = [Wg, f]$$

and since WH is dense in $\mathcal{H}$, it follows that $f = 0 \in \mathcal{H}$. Thus, 0 is not an eigenvalue of $W : \mathcal{H} \to \mathcal{H}$. This, together with the fact that $0 \leq [f, Wf]$ $\forall f \in \mathcal{H}$ implies that W is positive definite on $\mathcal{H}$.

(iii) Suppose $f \in V\mathcal{H}$. Then $f = Vg$ for some $g \in \mathcal{H}$. Since H is dense in there exists a sequence $\{g_n\}_{n=1}^{\infty} \subset H$ such that $|||g_n - g||| \to 0$ as $n \to \infty$. Let $f_n = Vg_n \in H$. Then for any $\varepsilon > 0$

$$\begin{aligned} \|f_m - f_n\|^2 &= \|V(g_m - g_n)\|^2 \\ &= [V(g_m - g_n), g_m - g_n] \\ &\leq |||V|||\ |||g_m - g_n|||^2 < \varepsilon \text{ for } m,n \text{ sufficiently large.} \end{aligned}$$

Thus, $\{f_n\}_{n=1}^{\infty}$ is a Cauchy sequence in H and since H is complete, $f_n \overset{H}{\to} h \in H$ as $n \to \infty$. Then

$$|||f_n - h|||^2 = (f_n - h, V(f_n - h)) \leq \|V\|\ \|f_n - h\|^2$$

so that $f_n \overset{\mathcal{H}}{\to} h$ as $n \to \infty$. However,

$$f_n = Vg_n \overset{\mathcal{H}}{\to} Vg = f.$$

Thus, $f = h \in H$ and the result is proved. □

Lemma 2.3: Let $T : H \supset \mathcal{D}(T) \to H$ be a closed operator with $0 \in \rho(T)$, the resolvent set of T, and let $V : H \to H$ be a positive definite bounded linear operator. Then

(i) $\mathcal{R}(V^{-1}T) = \mathcal{H}$ ($\mathcal{H}$ is defined in Lemma 2.2, and V is assumed to be extended to all of $\mathcal{H}$),

(ii) $\mathcal{D}(V^{-1}T)$ is dense in $\mathcal{H}$ and $(V^{-1}T)^{\#} = V^{-1}T^*$, where $*$ and $\#$ denote adjoints in H and $\mathcal{H}$ respectively,

(iii) $\mathcal{D}(V^{-1}T)$ is a core of T.

Proof: (i) By our assumptions $\mathcal{R}(T) = H$. By Lemma 2.2(iii), $\mathcal{D}(V^{-1}) \subset H$, so that $\mathcal{D}(V^{-1}) \subset \mathcal{R}(T)$. But $\mathcal{R}(V^{-1}) = \mathcal{H}$, so that $\mathcal{R}(V^{-1}T) = \mathcal{H}$.

(ii) Let $f \in \mathcal{D}(T)$. Then $Tf \in H$, and since $\mathcal{D}(V^{-1})$ is dense in H, there exists a sequence $\{g_n\}_{n=1}^{\infty} \subset \mathcal{D}(V^{-1})$ such that $g_n \overset{H}{\to} Tf$ as $n \to \infty$. By assumption $T^{-1} \in B(H)$, so that $f_n := T^{-1}g_n \overset{H}{\to} f$ as $n \to \infty$, and $f_n \in \mathcal{D}(V^{-1}T)$. Thus $\mathcal{D}(V^{-1}T)$ is dense in $\mathcal{D}(T)$ and so dense in H. It follows that $\mathcal{D}(V^{-1}T)$ is also dense in $\mathcal{H}$. Thus $(V^{-1}T)^{\#}$ can be defined.

Now, for $f \in \mathcal{D}(V^{-1}T^*)$ and $g \in \mathcal{D}(V^{-1}T)$

$$[f, V^{-1}Tg] = (f, Tg) = (T^*f, g) = [V^{-1}T^*f, g],$$

so that

$$V^{-1}T^* \subset (V^{-1}T)^{\#}. \tag{2.1}$$

Now suppose $f \in \mathcal{D}((V^{-1}T)^{\#})$ and $(V^{-1}T)^{\#}f = g$. Then

$$[f, V^{-1}Th] = [g, h] \ \forall h \in \mathcal{D}(V^{-1}T). \tag{2.2}$$

Since $\mathcal{R}(V^{-1}) = \mathcal{H}$, $g = V^{-1}v$ for some $v \in \mathcal{D}(V^{-1}) \subset H$. Further $\mathcal{R}(T^*) = H$, so that $v = T^*w$ for some $w \in \mathcal{D}(T^*)$. Thus (2.2) gives

$$[f, V^{-1}Th] = [V^{-1}T^*w, h] \ \forall h \in \mathcal{D}(V^{-1}T)$$

$$\Rightarrow [f, V^{-1}Th] = [w, V^{-1}Th] \ \ \forall h \in \mathcal{D}(V^{-1}T). \tag{2.3}$$

But by (i), $\mathcal{R}(V^{-1}T) = \mathcal{H}$, so that (2.3) implies that $w = f$, so that

$$(V^{-1}T)^{\#}f = V^{-1}T^*f,$$

i.e. $$(V^{-1}T)^{\#} \subset V^{-1}T^*, \tag{2.4}$$

and the result follows from (2.2) and (2.4).

(iii) Let $f \in \mathcal{D}(T)$. As in (ii) there exists a sequence $\{f_n\}_{n=1}^{\infty} \subset \mathcal{D}(V^{-1}T)$ such that $f_n \overset{H}{\to} f$, and $Tf_n \overset{H}{\to} Tf$ as $n \to \infty$. Hence $f \in \mathcal{D}\left(\overline{T|_{\mathcal{D}(V^{-1}T)}}\right)$. Since T is closed, the result follows. □

A more powerful result when T is self-adjoint is the following:

Lemma 2.4: Let $T : H \supset \mathcal{D}(T) \to H$ be self-adjoint with spectral family $\{E(\lambda) | \lambda \in \mathbb{R}\}$, and let $V \in B(H)$, $V > 0$. Assume there exists an open interval $I \ni 0$ and a constant $c > 0$, such that

$$\|E(I)f\| \leq c|||E(I)f||| \quad \forall f \in H. \tag{2.5}$$

Then $V^{-1}T$ is self-adjoint in $\mathcal{H}$.

Note: Here, we are using the notation $E(\cdot)$ to denote both the spectral family and the associated spectral measure of a self-adjoint operator, the argument being a real number in the former case and a set in the latter.

Proof: Firstly we note that (2.5) will hold with I replaced by any subinterval $I' \subset I$ since $E(I)E(I') = E(I')$.

Let $M = E(I)H$.

Since $E(I)$ is a projection, M is a closed subspace of H. For $u \in M$, (2.5) gives

$$\|u\| \leq c|||u|||$$

so that, if $\{u_n\}_{n=1}^{\infty} \subset M$ is a Cauchy sequence in $\mathcal{H}$, then it is a Cauchy sequence in H and $u_n \overset{H}{\to} u \in M$ as $n \to \infty$. Then $u_n \overset{\mathcal{H}}{\to} u$ and thus M is closed in $\mathcal{H}$. Let N and $\mathcal{N}$ be the orthogonal complements of M in H and $\mathcal{H}$ respectively, so that

$$H = M \oplus N \text{ and } \mathcal{H} = M \oplus \mathcal{N}.$$

Firstly, we prove that

$$V^{-1}(N \cap \mathcal{D}(V^{-1})) = N \tag{2.6}$$

(remember, V is assumed to be extended to all of H).

Let $f \in N$. Then

$$[f,u] = 0 \quad \forall u \in M$$

$$\Rightarrow (Vf,u) = 0 \; \forall u \in M$$

$$\Rightarrow Vf \in N,$$

i.e. $VN \subset N \cap \mathcal{D}(V^{-1})$ or $N \subset V^{-1}(N \cap \mathcal{D}(V^{-1}))$.

Conversely, suppose $f \in N \cap \mathcal{D}(V^{-1})$. Then, for all $u \in M$

$$[V^{-1}f,u] = (f,u) = 0$$

$$\Rightarrow V^{-1}f \in N,$$

i.e. $V^{-1}(N \cap \mathcal{D}(V^{-1})) \subset N,$

and the result (2.6) follows.

Now we show

$$N \cap \mathcal{D}(V^{-1}) \text{ is dense in } N. \tag{2.7}$$

Since V is positive definite and therefore one-to-one, it follows that

$$V^{-1} : N \cap \mathcal{D}(V^{-1}) \to N \text{ is a bijection.} \tag{2.8}$$

Let $f \in N$ be orthogonal to $N \cap \mathcal{D}(V^{-1})$. Then, $\forall u \in N \cap \mathcal{D}(V^{-1})$

$$0 = (f,u) = (f,VV^{-1}u) = [f,V^{-1}u]$$

and (2.8) implies that $f \in M$, so that $f = 0 \in H$, and the result (2.7) is proved.

Now we define operators

$$T(I) := TE(\mathbb{R}\backslash I) = \int_{\mathbb{R}\backslash I} \lambda dE(\lambda)$$

$$R(I) := \int_{\mathbb{R}\backslash I} \frac{1}{\lambda}\, dE(\lambda).$$

By the spectral theorem, $R(I)$ is bounded and self-adjoint and, moreover, $R(I)|_N$ is the inverse of $T|_N = T(I)|_N$. $N \cap \mathcal{D}(T)$ is dense in N since, if $f \in N$, then there exists a sequence $\{f_n\}_{n=1}^{\infty} \subset \mathcal{D}(T)$ such that $f_n \to f$ as $n \to \infty$, and so

$$E(\mathbb{R}\backslash I)f_n \in N \quad \text{and} \quad E(\mathbb{R}\backslash I)f_n \overset{H}{\to} E(\mathbb{R}\backslash I)f = f \text{ as } n \to \infty.$$

Further it is clear that

$$T(N \cap \mathcal{D}(T)) = T(I)\ (N \cap \mathcal{D}(T)) = N.$$

From (2.7) and the continuity of $R(I)$ it follows that

$$R(I)\ (N \cap \mathcal{D}(V^{-1})) \text{ is dense in } R(I)\ N = N \cap \mathcal{D}(T). \tag{2.9}$$

Thus

$$\begin{aligned} N \cap \mathcal{D}(V^{-1}T) &= \{u \in H \mid u \in N \cap \mathcal{D}(T),\quad Tu \in \mathcal{D}(V^{-1})\} \\ &= \{u \in H \mid u \in N \cap \mathcal{D}(T),\quad Tu \in N \cap \mathcal{D}(V^{-1})\} \\ &= \{u \in H \mid u \in N \cap \mathcal{D}(T),\quad u \in R(I)(N \cap \mathcal{D}(V^{-1}))\} \end{aligned}$$

i.e. $N \cap \mathcal{D}(V^{-1}T) = R(I)(N \cap \mathcal{D}(V^{-1}))$ (by (2.9)). (2.10)

Thus, $N \cap \mathcal{D}(V^{-1}T)$ is dense in $N \cap \mathcal{D}(T)$, and so is dense in N.

Now, we allow the length of the interval I, which will be denoted by $|I|$, to become arbitrarily small. Choose a sequence of intervals $\{I_n\}_{n=1}^{\infty}$, $0 \in I_n \subset I$ with $|I_n| \to 0$ as $n \to \infty$, and denote the corresponding set $(E(I_n)H)^{\perp}$ by N_n. We shall show that $\mathcal{D}(V^{-1}T)$ is dense in H.

Let $f \in H$. Then

$$f = E(\{0\})f + (Id - E(\{0\}))f$$

and

$$E(\{0\})f \in \mathcal{N}(T) \subset \mathcal{D}(V^{-1}T).$$

Further

$$f_n := E(\mathbb{R}\backslash I_n)f \in N_n \quad \text{and} \quad f_n \overset{H}{\to} (Id - E(\{0\}))f \quad \text{as} \quad n \to \infty.$$

But $N_n \cap \mathcal{D}(V^{-1}T)$ is dense in N_n, so there is a sequence

$$\{f_{nm}\}_{m=1}^{\infty} \subset N_n \cap \mathcal{D}(V^{-1}T) \; : \; f_{nm} \overset{H}{\to} f_n \quad \text{as} \quad m \to \infty.$$

Then, there is a 'diagonal sequence' $\{\tilde{f}_n\}_{n=1}^{\infty}=\{f_{nm(n)}\}_{n=1}^{\infty}$ such that

$$\{E(0)f + \tilde{f}_n\}_{n=1}^{\infty} \subset \mathcal{D}(V^{-1}T)$$

and

$$E(0)f + \tilde{f}_n \overset{H}{\to} f \quad \text{as} \quad n \to \infty,$$

which shows that $\mathcal{D}(V^{-1}T)$ is dense in H.

It is easy to show that $V^{-1}T$ is symmetric in H,

i.e. $V^{-1}T \subset (V^{-1}T)^{\#}$,

so it remains to show that $(V^{-1}T)^{\#} \subset V^{-1}T$.

Let $f \in \mathcal{D}((V^{-1}T)^{\#})$. Then, $\forall u \in \mathcal{D}(V^{-1}T)$ and some $g \in H$,

$$[f,V^{-1}Tu] = [g,u] = (Vg,u) = (h,u) \text{ where } h = Vg. \tag{2.11}$$

In particular, choose $u \in N \cap \mathcal{D}(V^{-1}T)$. Then

$$R(I)T(I)u = u$$

and so, from (2.11),

$$\begin{aligned}[f,V^{-1}Tu] &= (h,R(I)T(I)u) \\ &= (h,R(I)VV^{-1}T(I)u) \\ &= [R(I)h,V^{-1}Tu] \ \forall u \in N \cap \mathcal{D}(V^{-1}T).\end{aligned}$$

Therefore

$$[f - R(I)h,\ V^{-1}Tu] = 0 \quad \forall u \in N \cap \mathcal{D}(V^{-1}T). \tag{2.12}$$

From (2.10), $N \cap \mathcal{D}(V^{-1}T) = R(I)(N \cap \mathcal{D}(V^{-1}))$, so that

$$V^{-1}T(N \cap \mathcal{D}(V^{-1}T)) = V^{-1}T\ R(I)(N \cap \mathcal{D}(V^{-1}))$$

$$\Rightarrow V^{-1}\ (N \cap \mathcal{D}(V^{-1})) = \mathcal{N}, \quad \text{from (2.6).}$$

Thus, from (2.12)

$$[f - R(I)h,v] = 0 \ \forall v \in \mathcal{N}$$

$$\Rightarrow f - R(I)h \in M = \mathcal{N}(T(I)).$$

Thus, $f \in \mathcal{D}(T(I))$ and

$$T(I)[f - R(I)h] = T(I)f - E(\mathbb{R}\setminus I)h = 0$$

$$\Rightarrow TE(\mathbb{R}\setminus I)f = E(\mathbb{R}\setminus I)h,$$

i.e. $E(\mathbb{R}\setminus I)f \in \mathcal{D}(T)$.

But $E(I)f \in \mathcal{D}(T)$ so that $f = E(I)f + E(\mathbb{R}\setminus I)f \in \mathcal{D}(T)$. Then

$$[f,V^{-1}Tu] = (f,Tu) = (Tf,u) \ \forall u \in \mathcal{D}(V^{-1}T). \tag{2.13}$$

From (2.11) and (2.13) we obtain

$$Tf = h = Vg$$
$$\Rightarrow g = V^{-1}Tf,$$
i.e. $\quad g := (V^{-1}T)^{\#}f = V^{-1}Tf$

and so

$$(V^{-1}T)^{\#} \subset V^{-1}T.$$

Thus, $V^{-1}T\colon \mathcal{H} \supset \mathcal{D}(V^{-1}T) \to \mathcal{H}$ is self-adjoint. $\square$

Lemma 2.5: Under the same assumptions at the previous lemma, there exist constants $a,b > 0$ such that

$$\|u\| \leq a\|Tu\| + b|||u||| \quad \forall u \in \mathcal{D}(T) \tag{2.14}$$

and then

$$\|u\| \leq \sqrt{2}\ \mathrm{Max}\ (a\|V\|^{1/2},b)\ |||(V^{-1}T\pm i)u||| \quad \forall u \in \mathcal{D}(V^{-1}T). \tag{2.15}$$

Proof:

$$\begin{aligned}\|Tu\|^2 &= (Tu,Tu)\\ &= (Tu,VV^{-1}Tu) \qquad\qquad \forall u \in \mathcal{D}(V^{-1}T)\\ &= [Tu,V^{-1}Tu]\\ &\leq |||Tu|||\ |||V^{-1}Tu|||\\ &\leq (Tu,VTu)^{1/2}|||V^{-1}Tu|||\\ &\leq \|V\|^{1/2}\|Tu\|\ \ |||V^{-1}Tu|||.\end{aligned}$$

Thus

$$\|Tu\| \leq \|V\|^{1/2}|||V^{-1}Tu|||.$$

Then, if (2.14) holds

$$\|u\| \leq a\|V\|^{1/2}|||V^{-1}Tu||| + b|||u|||$$

$$\leq \text{Max}\,(a\|V\|^{1/2},b)(|||V^{-1}Tu||| + |||u|||)$$

$$\leq \sqrt{2}\,\text{Max}\,(a\|V\|^{1/2},b)\,|||(V^{-1}T \pm i)u||| \quad \forall u \in \mathcal{D}(V^{-1}T).$$

It remains to prove (2.14). Let $\{P(\tau) | \tau \in \mathbb{R}\}$ be the spectral family of V in H, and let $I \subset \mathbb{R}$ be an open neighbourhood of zero. Then, for $\tau > 0$,

$$\begin{aligned}\|(Id - P(\tau)E(I))u\| &\leq \|(Id - E(I))u\| + \|(Id - P(\tau))E(I)u\| \\ &= \|(Id - E(I))u\| + \|(Id - P(\tau))(E(I) - P(\tau))u\| \\ &\leq \|(Id - E(I))u\| + \|(E(I) - P(\tau))u\| \\ &\leq 2\|(Id - E(I))u\| + \|(Id - P(\tau))u\|. \end{aligned} \tag{2.16}$$

Now

$$\begin{aligned}\|(Id - P(\tau))u\|^2 &= ((Id - P(\tau))u, \ (Id - P(\tau))u) \\ &\leq \tau^{-1}((Id - P(\tau))u, V(Id - P(\tau))u) \\ &= \tau^{-1}|||(Id - P(\tau))u|||^2;\end{aligned}$$

here we have used the spectral theorem. Thus

$$\|(Id - P(\tau))u\| \leq \tau^{-1/2}|||(Id - P(\tau))u||| \leq \tau^{-1/2}|||u|||. \tag{2.17}$$

Again using the spectral theorem,

$$\|(Id - E(I))u\| \leq m^{-1}\|T(Id - E(I))u\| \leq m^{-1}\|Tu\| \quad \forall u \in \mathcal{D}(T) \tag{2.18}$$

where

$$m = \min(\lambda_1,\lambda_2) \text{ with } I = (-\lambda_1,\lambda_2).$$

(2.16) - (2.18) imply that

$$\|(Id - P(\tau)E(I))u\| \leq 2m^{-1}\|Tu\| + \tau^{-1/2}|||u||| \quad \forall u \in \mathcal{D}(T). \tag{2.19}$$

From (2.5)

$$\|E(I)u\|^2 \leq c^2(E(I)u, V\,E(I)u)$$

$$= c^2(V^{1/2}E(I)u, V^{1/2}E(I)u)$$

$$= c^2\|V^{1/2}E(I)u\|^2,$$

i.e. $\|E(I)u\|^2 - c^2\|V^{1/2}E(I)u\|^2 \leq 0.$

Hence

$$\int_0^{\|V\|} (1 - \tau c^2)d\|P(\tau)E(I)u\|^2$$

$$= \|E(I)u\|^2 - c^2\|V^{1/2}E(I)u\|^2 \leq 0.$$

Now, for any $\alpha > 1$

$$1 - \tau c^2 \geq \begin{cases} 1 - \alpha^{-1} & \text{for } 0 < \tau \leq (\alpha c^2)^{-1} =: \varepsilon, \\ 1 - \|V\|c^2 & \text{for } \varepsilon \leq \tau \leq \|V\|. \end{cases}$$

Thus, for any $\alpha > 1$

$$(1 - \alpha^{-1})\|P(\varepsilon)E(I)u\|^2 + (1 - \|V\|c^2)\|(Id - P(\varepsilon))E(I)u\|^2$$

$$\leq \int_0^{\|V\|} (1 - \tau c^2)d\|P(\tau)E(I)u\|^2 \leq 0,$$

i.e. $(1 - \alpha^{-1})\|P(\varepsilon)E(I)u\|^2 \leq (\|V\|c^2 - 1)\|(Id - P(\varepsilon))E(I)u\|^2.$

Hence

$$\|V\|c^2\|P(\varepsilon)E(I)u\|^2 = (\|V\|c^2 - 1)\|P(\varepsilon)E(I)u\|^2 + \|P(\varepsilon)E(I)u\|^2$$

$$\leq (\|V\|c^2 - 1)\|P(\varepsilon)E(I)u\|^2 + \frac{\alpha}{\alpha-1}(\|V\|c^2 - 1)\|(Id - P(\varepsilon))E(I)u\|^2$$

$$\leq \frac{\alpha}{\alpha-1}(\|V\|c^2 - 1)\|E(I)u\|^2$$

$$\leq \frac{\alpha}{\alpha-1}(\|V\|c^2 - 1)\|u\|^2.$$

Therefore, for any particular choice of $\alpha > 1$

$$\|P(\varepsilon)E(I)u\| \leq \sqrt{\frac{\alpha}{\alpha-1}}\sqrt{1 - (\|V\|c^2)^{-1}}\ \|u\| =: K\|u\|,$$

where $0 < K < 1$, so that

$$\|(\mathrm{Id} - P(\varepsilon)E(I))u\| \geq \|u\| - \|P(\varepsilon)E(I)u\| \geq (1 - K)\|u\| =: K'\|u\| \qquad (2.20)$$

where $0 < K' < 1$.

Finally using (2.19) and (2.20) we obtain

$$\|u\| \leq \frac{2}{mK'}\|Tu\| + \frac{1}{\varepsilon^{1/2}K'}\ |||u||| \quad \forall u \in \mathcal{D}(T)$$

which is (2.14). Note that the coefficients $a = \frac{2}{mK'}$, $b = \frac{1}{\varepsilon^{1/2}K'}$ depend only on the interval $I = (-\lambda_1,\lambda_2)$ and the constant c. □

Now, let $V_s \in B(H)$, $V_s > 0$, $s = 1,2$, and

$$V_1 + V_2 = \mathrm{Id}. \qquad (2.21)$$

V_1 and V_2 commute, and, as in Lemma 2.2, we define

$$[.,.]_s := (.,V_s.)\ , \qquad |||\cdot|||_s := \sqrt{[.,.]_s}$$

$$\mathcal{H}_s := \overline{\{H,[.,.]_s\}}, \qquad s = 1,2; \qquad (2.22)$$

(i.e. $\mathcal{H}_s$ is the completion of H with respect to the inner product $[.,.]_s$).

By Lemma 2.2(ii) V_1 and V_2 can be extended to positive definite operators on both $\mathcal{H}_1$ and $\mathcal{H}_2$. We use the same symbols for the extended operators.

Adjoints in $\mathcal{H}_s$ are denoted by $^{\#s}$, $s = 1,2$.

The next two lemmas give some estimates for obtaining inverses of various operators:

Lemma 2.6: Let $T : H \supset \mathcal{D}(T) \to H$ be self-adjoint. Then, for $\lambda,\mu \in \mathbb{C}$ and $f \in \mathcal{D}(V_1^{-1}(T-\lambda))$,

$$|||(V_1^{-1}(T-\lambda V_2)-\mu)f|||_1^2 \geq (\text{Im } \mu)^2 |||f|||_1^2 + 2 \text{ Im } \mu \text{ Im } \lambda |||f|||_2^2. \tag{2.23}$$

Then, provided $\text{Im } \lambda \text{ Im } \mu > 0$,

(i) $(V_1^{-1}(T-\lambda V_2)-\mu)\mathcal{D}(V_1^{-1}(T-\lambda)) = H_1$;

(ii) $(V_1^{-1}(T-\lambda V_2)-\mu)$ is one-to-one;

(iii) $|||(V_1^{-1}(T-\lambda V_2)-\mu)^{-1}g|||_1 \leq \dfrac{1}{|\text{Im } \mu|} |||g|||_1, \ \forall g \in H_1$;

$$|||(V_1^{-1}(T-\lambda V_2)-\mu)^{-1}g|||_2 \leq \frac{1}{\sqrt{2 \text{ Im } \mu \text{ Im } \lambda}} |||g|||_1, \ \forall g \in H_1. \tag{2.24}$$

Proof: Since $V_1 + V_2 = \text{Id}$, $V_1^{-1}(T-\lambda V_2)-\mu = V_1^{-1}(T-\lambda)-\mu+\lambda$ and, clearly, $\mathcal{D}(V_1^{-1}(T-\lambda V_2)) = \mathcal{D}(V_1^{-1}(T-\lambda))$. Then, for $f \in \mathcal{D}(V_1^{-1}(T-\lambda))$,

$$|||(V_1^{-1}(T-\lambda V_2)-\mu)f|||_1^2 = |||((V_1^{-1}(T-\lambda V_2) - \text{Re } \mu) - i \text{ Im } \mu)f|||_1^2$$

$$= |||(V_1^{-1}(T-\lambda V_2) - \text{Re } \mu)f|||_1^2 + (\text{Im } \mu)^2 \ |||f|||_1^2$$

$$- i \text{ Im } \mu[(V_1^{-1}(T-\lambda V_2) - \text{Re } \mu)f,f]_1 + i \text{ Im } \mu[f,(V_1^{-1}(T-\lambda V_2)-\text{Re } \mu)f]_1$$

$$\geq (\text{Im } \mu)^2 |||f|||_1^2 - i \text{ Im } \mu\{((T-\lambda V_2-(\text{Re } \mu)V_1)f,f)-(f,(T-\lambda V_2-(\text{Re } \mu)V_1)f)\}$$

$$= (\text{Im } \mu)^2 |||f|||_1^2 - i \text{ Im } \mu\{(f,(T-\bar{\lambda}V_2-(\text{Re } \mu)V_1)f)-(f,T-\lambda V_2-(\text{Re } \mu)V_1f)\}$$

$$= (\text{Im } \mu)^2 |||f|||_1^2 - i \text{ Im } \mu\{(f,(\lambda-\bar{\lambda})V_2f)\}$$

$$= (\text{Im } \mu)^2 |||f|||_1^2 + 2 \text{ Im } \mu \text{ Im } \lambda(f,V_2f)$$

$$= (\text{Im } \mu)^2 |||f|||_1^2 + 2 \text{ Im } \mu \text{ Im } \lambda |||f|||_2^2$$

which proves (2.23).

Now, if $\operatorname{Im}\mu \operatorname{Im}\lambda > 0$, it follows easily from (2.23) that (ii) holds and the estimates (iii) follow immediately, at least for $g \in \mathcal{R}(V_1^{-1}(T-\lambda V_2)-\mu)$. We need prove only that this range is all of $\mathcal{H}_1$.

Now, $\operatorname{Im}\mu \operatorname{Im}\lambda > 0$ implies that $\operatorname{Im}\overline{\lambda} \neq 0$, so that, since T is self-adjoint, $(T-\overline{\lambda})^{-1} \in B(H)$ and, by Lemma 2.3,

$$V_1^{-1}(T-\overline{\lambda}))^{\#1} = V_1^{-1}(T-\lambda)$$

Then, since $(\overline{\mu} - \overline{\lambda})\mathrm{Id}$ is bounded on $\mathcal{H}_s$,

$$(V_1^{-1}(T - \overline{\lambda}V_2)-\overline{\mu}))^{\#1} = (V_1^{-1}(T-\overline{\lambda}) - (\overline{\mu}-\overline{\lambda}))^{\#1} = (V_1^{-1}(T-\overline{\lambda}))^{\#1} - ((\overline{\mu}-\overline{\lambda})\mathrm{Id})^{\#1}$$

$$= V_1^{-1}(T-\lambda) - (\mu-\lambda) = V_1^{-1}(T-\lambda V_2)-\mu$$

so that $V_1^{-1}(T-\lambda V_2)-\mu$ is a closed operator in $\mathcal{H}_1$.

Further, (2.23) with $\overline{\lambda}$ and $\overline{\mu}$ in place of λ and μ, shows that

$$\{0\} = \mathcal{N}(V_1^{-1}(T-\overline{\lambda}V_2)-\overline{\mu}) = \mathcal{R}(V_1^{-1}(T-\lambda V_2)-\mu)^{\perp}$$

so that $\mathcal{R}(V_1^{-1}(T-\lambda V_2)-\mu)$ is dense in $\mathcal{H}_1$.

Now we have that $(V_1^{-1}(T-\lambda V_2)-\mu)^{-1}$ is a closed, densely defined and bounded operator in $\mathcal{H}_1$ which implies that

$$\mathcal{D}((V_1^{-1}(T-\lambda V_2)-\mu)^{-1}) = \mathcal{R}(V_1^{-1}(T-\lambda V_2)-\mu) = \mathcal{H}_1$$

and the proof is complete. □

<u>Lemma 2.7:</u> Let $T : H \supset \mathcal{D}(T) \to H$ be self-adjoint. Then for $f \in \mathcal{D}(T)$, and $\lambda,\mu \in \mathbb{C}$,

$$\|(T-\lambda V_1-\mu V_2)f\|^2 \geq \begin{cases} (\operatorname{Im}\lambda)^2\|f\|^2 + 2\operatorname{Im}\lambda(\operatorname{Im}\mu - \operatorname{Im}\lambda)|||f|||_2^2 \\ \\ (\operatorname{Im}\mu)^2\|f\|^2 + 2\operatorname{Im}\mu(\operatorname{Im}\lambda - \operatorname{Im}\mu)|||f|||_1^2 \end{cases} \qquad (2.25)$$

Further, if $\operatorname{Im}\lambda \operatorname{Im}\mu > 0$, then $R(\lambda,\mu) := (T-\lambda V_1-\mu V_2)^{-1} \in B(H)$ and

$$\|R(\lambda,\mu)g\|^2 \leq \operatorname{Max}\left(\frac{1}{(\operatorname{Im}\lambda)^2}, \frac{1}{(\operatorname{Im}\mu)^2}\right)\|g\|^2 \quad \forall g \in H. \tag{2.26}$$

<u>Proof:</u> For $f \in \mathcal{D}(T)$

$$\|(T-\lambda V_1-\mu V_2)f\|^2 = \begin{cases} \|(T-\lambda-(\mu-\lambda)V_2)f\|^2 \\ \|(T-(\lambda-\mu)V_1-\mu)f\|^2 \end{cases}$$

$$= \begin{cases} \|(T-(\mu-\lambda)V_2-\operatorname{Re}\lambda)f-i(\operatorname{Im}\lambda)f\|^2 \\ \|(T-(\lambda-\mu)V_1-\operatorname{Re}\mu)f-i(\operatorname{Im}\mu)f\|^2 \end{cases}$$

$$\geq \begin{cases} (\operatorname{Im}\lambda)^2\|f\|^2 + 2\operatorname{Im}\lambda(\operatorname{Im}\mu-\operatorname{Im}\lambda)\ |||f|||_2^2 \\ (\operatorname{Im}\mu)^2\|f\|^2 + 2\operatorname{Im}\mu(\operatorname{Im}\lambda-\operatorname{Im}\mu)\ |||f|||_1^2. \end{cases}$$

which proves (2.25).

Now, if $\operatorname{Im}\lambda \cdot \operatorname{Im}\mu > 0$ then either $\operatorname{Im}\lambda\ (\operatorname{Im}\mu-\operatorname{Im}\lambda) \geq 0$ or $\operatorname{Im}\mu\ (\operatorname{Im}\lambda-\operatorname{Im}\mu) \geq 0$ so that (2.25) show that $T-\lambda V_1-\mu V_2$ is one-to-one and (2.26) follows at least for g in $\mathcal{R}(T-\lambda V_1-\mu V_2)$. That this range is all of H can be proved as in the previous lemma. □

The next three lemmas relate the existence of the resolvent $R(\lambda_1,\lambda_2) \in B(H)$ to conditions on the operator T of the form (2.5). $T: H \supset \mathcal{D}(T) \to H$ is always self-adjoint.

<u>Lemma 2.8:</u> Let $R(i,\lambda_o) = (T-iV_1-\lambda_o V_2)^{-1} \in B(H)$. Then for all $\alpha \in \mathbb{R}$, $T-\alpha V_1$ is self-adjoint; let $\{E^\alpha(\lambda)\,|\,\lambda \in \mathbb{R}\}$ be the spectral family of $T-\alpha V_1$. Then, for all $\alpha \in \mathbb{R}$, there exists an open interval $I \ni \lambda_o$ and a constant $c > 0$ such that

$$\|E^\alpha(I)u\| \leq c\,|||E^\alpha(I)u|||_1 \quad \forall u \in H. \tag{2.27}$$

Moreover, if α is restricted to $K \subset\subset \mathbb{R}$, then c and I can be chosen independently of α.

Proof: Assume that the result is false, i.e. that there do not exist $c > 0$ and $I \ni \lambda_o$ such that (2.27) holds for all $\alpha \in K \subset\subset \mathbb{R}$. Then, choose sequences

$$\{\lambda_n'\}_{n=1}^{\infty} \text{ converging to } \lambda_o, \ \{\lambda_n''\}_{n=1}^{\infty} \text{ converging to } \lambda_o \text{ as } n \to \infty$$

such that

$$\lambda_n' < \lambda_o < \lambda_n'' \quad \forall n$$

and let $I_n = (\lambda_n', \lambda_n'')$. For each n, choose $u_n \in H$ and $\alpha_n \in K$ such that

$$|||E^{\alpha_n}(I_n)u_n|||_1 < \frac{1}{n} \quad \text{and} \quad \|E^{\alpha_n}(I_n)u_n\| = 1$$

Let $v_n = E^{\alpha_n}(I_n)u_n$. Then

$$\|V_1^{1/2}v_n\|^2 = |||v_n|||_1^2 < \frac{1}{n^2}$$

so that

$$\|V_1^{1/2}v_n\| \to 0 \quad \text{as} \quad n \to \infty$$

$$\Rightarrow \|V_1 \ v_n\| \to 0 \quad \text{as} \quad n \to \infty. \tag{2.28}$$

Further

$$(T-\alpha_n V_1-\lambda_o)v_n = \int_{I_n} (\lambda-\lambda_o)dE^{\alpha_n}(\lambda)v_n \to 0 \in H \quad \text{as} \quad n \to \infty \tag{2.29}$$

since $\alpha_n \in K$ for all n.

But $\|v_n\| = 1 \ \forall n$ and from (2.28) and (2.29)

$$(T-iV_1-\lambda_o V_2)v_n = (T-\alpha_n V_1-\lambda_o)v_n + (\alpha_n - i+\lambda_o)V_1 v_n$$

$$\to 0 \in H \quad \text{as} \quad n \to \infty$$

and this contradicts the boundedness of $R(i,\lambda_o)$. □

To derive a near converse of Lemma 2.8 we require

Lemma 2.9: Let $\{E(\lambda)|\lambda \in \mathbb{R}\}$ be the spectral family of T. Suppose that, for some $\lambda_o \in \mathbb{R}$ there exists an open interval $I \ni \lambda_o$ and $c > 0$ such that

$$\|E(I)u\| \leq c|||E(I)u|||_1 \quad \forall u \in H.$$

Then, for $R_1(\pm i,\lambda_o) := (V_1^{-1}(T-\lambda_o V_2)\mp i)^{-1}$, there exists $c' > 0$ such that

$$\|R_1(\pm i,\lambda_o)u\| \leq c'|||u|||_1 \quad \forall u \in H_1.$$

Proof: The assumptions of Lemma 2.4 hold for $T-\lambda_o$. Thus, $V_1^{-1}(T-\lambda_o) = V_1^{-1}(T-\lambda_o V_2)-\lambda_o$ is self-adjoint in H_1, so that $V_1^{-1}(T-\lambda_o V_2)$ is also self-adjoint in H_1. Therefore $(V_1^{-1}(T-\lambda_o V_2)\mp i)^{-1} \in B(H_1)$.

Now suppose the result is false. Choose a sequence $\{u_n\}_{n=1}^{\infty} \subset H_1$ such that $|||u_n|||_1 = 1$ and $\|R_1(\pm i,\lambda_o)u_n\| > n$.

Let $w_n = R_1(\pm i,\lambda_o)\left[\frac{u_n}{\|R_1(\pm i,\lambda_o)u_n\|}\right]$ so that $\|w_n\| = 1$. Then

$$|||V_1^{-1}(T-\lambda_o V_2)w_n|||_1^2 + |||w_n|||_1^2 = |||(V_1^{-1}(T-\lambda_o V_2)\mp i)w_n|||_1^2$$
$$= \frac{|||u_n|||_1^2}{\|R_1(\pm i,\lambda_o)u_n\|^2} < \frac{1}{n^2}$$

so that

$$w_n \overset{H_1}{\to} 0 \text{ as } n \to \infty \tag{2.30}$$

and

$$V_1^{-1}(T-\lambda_o)w_n = V_1^{-1}(T-\lambda_o V_2)w_n - \lambda_o w_n \overset{H_1}{\to} 0 \quad \text{as} \quad n \to \infty. \tag{2.31}$$

Since

$$\|V_1 u\|^2 = (V_1 u, V_1 u) = (V_1^{1/2}u, V_1^{3/2}u) \leq |||u|||_1^2 \quad \forall u \in H_1$$

we deduce from (2.31) that

$$(T-\lambda_o)w_n \overset{H}{\to} 0 \text{ as } n \to \infty.$$

Then, using Lemma 2.5 with $T-\lambda_o$ in place of T, we deduce that there exist constants a, b > 0 such that

$$\|w_n\| \leq a\|(T-\lambda_o)w_n\| + b|||w_n|||_1 \to 0 \quad \text{as} \quad n \to \infty$$

which contradicts $\|w_n\| = 1$. □

Now, the near converse of Lemma 2.8 is

Lemma 2.10: Let the assumptions of Lemma 2.9 hold. Then $R(\pm i,\lambda_o) \in B(H)$.

Proof: From Lemma 2.9 we have

$$\|R_1(\pm i,\lambda_o)u\| \leq c'|||u|||_1 \quad \forall u \in H_1.$$

Since $R_1(\pm i,\lambda_o)u = R(\pm i,\lambda_o)V_1u$, we see that $R(\pm i,\lambda_o)$ is defined on $R(V_1)$ and, for $v \in R(V_1)$, $u \in H_1$

$$[R(\pm i,\lambda_o)v,u]_1 = (R(\pm i,\lambda_o)v,V_1u) = (v,R(\mp i,\lambda_o)V_1u).$$

Therefore

$$|[R(\pm i,\lambda_o)v,u]_1| \leq \|v\|\|R(\mp i,\lambda_o)V_1u\|$$

$$\leq c'\|v\| \, |||u|||_1 .$$

Taking $u = R(\pm i,\lambda_o)v$, we see that

$$|||R(\pm i,\lambda_o)v||| \leq c'\|v\|.$$

Then, applying Lemma 2.5, there exist constants a, b > 0 such that

$$\|R(\pm i,\lambda_o)v\| \le a\|(T-\lambda_o)R(\pm i,\lambda_o)v\| + b|||R(\pm i,\lambda_o)v|||_1$$

$$\le a\|(T\mp iV_1-\lambda_o V_2)R(\pm i,\lambda_o)v\| + a\|(\pm iV_1+\lambda_o V_1)R(\pm i,\lambda_o)v\|$$

$$+ b|||R(\pm i,\lambda_o)v|||_1$$

$$\le (a + (a|\pm i+\lambda_o|+b)c')\|v\|$$

i.e. $R(\pm i,\lambda_o)$ is bounded on $\mathcal{R}(V_1)$ which is dense in H. Since T is closed, $T \pm iV_1 - \lambda_o V_2$ is closed so that $R(\pm i,\lambda_o)$ is also closed and therefore everywhere defined on H. □

The next lemma introduces an analytic function defined in terms of $R(\lambda,\mu)$.

<u>Lemma 2.11:</u> Let $(\lambda_o,\mu_o) \in \mathbb{C}^2$ be such that

$$R(\lambda_o,\mu_o) = (T-\lambda_o V_1-\mu_o V_2)^{-1} \in B(H),$$

so that there exists a constant $c > 0$ such that

$$\|R(\lambda_o,\mu_o)g\| \le c\|g\| \quad \forall g \in H.$$

Then $R(\lambda,\mu_o) \in B(H)$ for all $\lambda:|\lambda-\lambda_o| < \frac{1}{c}$ and

$$\|R(\lambda,\mu_o)g\| \le \frac{c}{1-c|\lambda-\lambda_o|}\|g\| \quad \forall g \in H.$$

Further, for $f,g \in H$

$$\varphi_{f,g}(\lambda) := (f,R(\lambda,\mu_o)g)$$

is an analytic function of λ in $\{\lambda \in \mathbb{C} | \, |\lambda-\lambda_o| < \frac{1}{c}\}$.

<u>Proof:</u> $W(\lambda,\mu_0) := T-\lambda V_1-\mu_0 V_2$

$$= T-\lambda_0 V_1-\mu_0 V_2-(\lambda-\lambda_0)V_1$$

$$= [Id-(\lambda-\lambda_0)V_1 R(\lambda_0,\mu_0)](T-\lambda_0 V_1-\mu_0 V_2)$$

$$= (Id-Q)W(\lambda_0,\mu_0)$$

where $Q := (\lambda-\lambda_0)V_1 R(\lambda_0,\mu_0)$. Thus, $R(\lambda,\mu_0) \in B(H)$ iff $(Id-Q)^{-1} \in B(H)$ and, by a standard result, this is true if $\|Q\| < 1$. Now

$$\|Q\| \le |\lambda-\lambda_0|\|V_1\|\|R(\lambda_0,\mu_0)\| \le |\lambda-\lambda_0|c,$$

so that it is sufficient that

$$|\lambda-\lambda_0| < \frac{1}{c} .$$

Again, by a standard result,

$$\|(Id-Q)^{-1}\| \le \frac{1}{1-\|Q\|} \le \frac{1}{1-|\lambda-\lambda_0|c} .$$

Then

$$\|R(\lambda,\mu_0)\| = \|R(\lambda_0,\mu_0)(Id-Q)^{-1}\|$$

$$\le \|R(\lambda_0,\mu_0)\|\|(Id-Q)^{-1}\| \le \frac{c}{1-|\lambda-\lambda_0|c} .$$

Since $(W(\lambda,\mu))^* = W(\overline{\lambda},\overline{\mu})$, it follows that

$$R(\overline{\lambda},\overline{\mu}_0) = [R(\lambda,\mu_0)]^*$$

and

$$\|R(\overline{\lambda},\overline{\mu}_0)\| \le \frac{c}{1-c|\lambda-\lambda_0|} \quad \text{for} \quad |\lambda-\lambda_0| < \frac{1}{c} .$$

Therefore, if in addition $|\lambda'-\lambda_0| < \frac{1}{c}$, we have a generalised resolvent equation

$$R(\lambda',\mu_0) - R(\lambda,\mu_0) = (\lambda'-\lambda)R(\lambda',\mu_0)V_1R(\lambda,\mu_0)$$

so that for any $f,g \in H$

$$\begin{aligned}\frac{(f,R(\lambda',\mu_0)g)-(f,R(\lambda,\mu_0)g)}{\lambda'-\lambda} &= (f,R(\lambda',\mu_0)V_1R(\lambda,\mu_0)g) \\ &= (R(\overline{\lambda}',\overline{\mu}_0)f,V_1R(\lambda,\mu_0)g).\end{aligned} \qquad (2.32)$$

Choose $\varepsilon > 0$ such that λ and λ' lie in the disc $|\lambda-\lambda_0| \leq \frac{1}{c} - \varepsilon$. Then

$$\begin{aligned}&|(f,R(\lambda',\mu_0)g) - (f,R(\lambda,\mu_0)g)| \\ &= |\lambda'-\lambda||(f,R(\lambda',\mu_0)V_1R(\lambda,\mu_0)g)| \\ &\leq |\lambda'-\lambda|\|f\|\|R(\lambda',\mu_0)\|\|V_1\|\|R(\lambda,\mu_0)\|\|g\| \\ &\leq a(\varepsilon)|\lambda'-\lambda|\|f\|\|g\|,\end{aligned}$$

where $a(\varepsilon)$ is a positive constant depending on the choice of ε. Therefore, as $\lambda' \to \lambda$

$$(f,R(\lambda',\mu_0)g) \to (f,R(\lambda,\mu_0)g)$$

and it follows easily that, for any $g \in H$

$$(R(\overline{\lambda}',\overline{\mu}_0)f,g) \to (R(\overline{\lambda},\overline{\mu}_0)f,g).$$

Therefore the right hand side of (2.32) converges to $(R(\overline{\lambda},\overline{\mu}_0)f,V_1R(\lambda,\mu_0)g)$ as $\lambda' \to \lambda$. This proves that the derivative of $\varphi_{f,g}(\lambda) := (f,R(\lambda,\mu_0)g)$ exists for all λ s.t. $|\lambda-\lambda_0| < \frac{1}{c}$. □

We end this section of preliminary results by introducing the idea of a totally orthogonal set.

<u>Definition 2.12:</u> A set of vectors $\{f_n\}_{n=1}^{\infty} \subset H$ is said to be totally orthogonal with respect to the positive definite operators V_1 and V_2 if

$$(f_m, f_n) = [f_m, f_n]_1 = [f_m, f_n]_2 = 0 \quad \text{for} \quad m \neq n.$$

<u>Lemma 2.13:</u> Any finite dimensional subspace M of H contains a complete totally orthogonal set of vectors.

<u>Proof:</u> Let $P_M : H \to H$ be the projection of H onto M. Then $V_1^M := P_M V_1 P_M$ is the restriction of V_1 to M and is positive definite on M equipped with the inner product $(.,.)$, and $\|V_1^M\| \leq 1$.

Thus V_1^M has a complete set of eigenvectors $\{f_1, f_2, \ldots, f_m\} \subset M (m = \dim M)$ and corresponding eigenvalues $0 < \lambda_1, \lambda_2, \ldots, \lambda_m < 1$. Now, eigenvectors are orthogonal and can be normalised so that we have

$$(f_i, f_j) = \delta_{ij},$$

$$[f_i, f_j]_1 = (f_i, V_1 f_j) = (f_i, \lambda_j f_j) = \lambda_j (f_i, f_j) = \lambda_j \delta_{ij},$$

$$[f_i, f_j]_2 = (f_i, f_j) - [f_i, f_j]_1 = (1-\lambda_j)\delta_{ij},$$

where δ_{ij} is the Kronecker delta. Thus, the set of eigenvectors $\{f_1, \ldots, f_m\}$ is totally orthogonal in M with respect to V_1 and V_2. □

3 The two-parameter system

Let H_r, $r = 1,2$, be separable Hilbert spaces with inner products $(.,.)_r$ and norms $\|\cdot\|_r := \sqrt{(.,.)_r}$, $r = 1,2$. Let $T_r : H_r \supset \mathcal{D}(T_r) \to H_r$, $r = 1,2$, be self-adjoint operators, and let $V_{rs} : H_r = \mathcal{D}(V_{rs}) \to H_r$, $r,s = 1,2$ be bounded self-adjoint operators satisfying

$$\begin{aligned} V_{11} &> 0, & V_{12} &> 0 \\ -V_{21} &> 0, & V_{22} &> 0 \end{aligned} \tag{3.1}$$

and

$$|V_{r1}| + V_{r2} = \text{Id}, \; r = 1,2, \tag{3.2}$$

where $|V_{rs}|$ denote the positive square root of V_{rs}^2.

By Lemma 2.2

$$[.,.]_{rs} := (.,|V_{rs}|.)_r \; , \qquad |||\cdot|||_{rs} = \sqrt{[.,.]}, \quad s = 1,2$$

define inner products and associated norms on H_r, $r = 1,2$. Then we define

$$\mathcal{H}_{rs} := \overline{\{H_r,[.,.]_{rs}\}} \; , \quad r,s = 1,2. \tag{3.3}$$

The two parameter system to be considered is

$$W_r(\underset{\sim}{\lambda}) := T_r-\lambda_1 V_{r1}-\lambda_2 V_{r2} : H_r \supset \mathcal{D}(T_r) \to H_r, \quad r = 1,2, \tag{3.4}$$

where $\underset{\sim}{\lambda} = (\lambda_1,\lambda_2) \in \mathbb{C}^2$ is a pair of spectral parameters linking the operators W_1 and W_2.

As is usual in multiparameter theory (See e.g. Sleeman [9]) we consider the system of operators induced by (3.4) in the tensor product space $H^{\otimes} = H_1 \otimes H_2$ which can be thought of as follows:

$$H_a^{\otimes} := \mathrm{Sp}[f_1 \otimes f_2 \mid f_r \in H_r, \quad r = 1,2],$$

with inner product $(.,.)^{\otimes}$, defined for separable elements $f_1 \otimes f_2$, $g_1 \otimes g_2 \in H_a^{\otimes}$ by

$$(f_1 \otimes f_2, g_1 \otimes g_2)^{\otimes} := (f_1, f_1)_1 (f_2, g_2)_2,$$

and extended by linearity to $H_a^{\otimes}$, and norm $\|\cdot\|^{\otimes} := \sqrt{(.,.)^{\otimes}}$. Then

$$H^{\otimes} := \overline{\{H_a^{\otimes}, (.,.)^{\otimes}\}}.$$

For later purposes we define here a dense subspace $\mathcal{D} \subset H$:

$$\mathcal{D} := \mathcal{D}(T_1) \otimes_a \mathcal{D}(T_2) := \mathrm{Sp}[f_1 \otimes f_2 \mid f_r \in \mathcal{D}(T_r), \quad r = 1,2]. \tag{3.5}$$

Induced operators in $H^{\otimes}$ are defined in the usual way, e.g.

$$\mathcal{D}(T_1 \otimes_a \mathrm{Id}) := \mathcal{D}(T_1) \otimes_a H_2 = \mathrm{Sp}[f_1 \otimes f_2 \mid f_1 \in \mathcal{D}(T_1), f_2 \in H_2]$$

$$(T_1 \otimes_a \mathrm{Id})(f_1 \otimes f_2) := T_1 f_1 \otimes f_2,$$

extended by linearity to $\mathcal{D}(T_1 \otimes_a \mathrm{Id})$; then

$$T_1^{\otimes} := \overline{T_1 \otimes_a \mathrm{Id}}, \text{ the closure of } T_1 \otimes_a \mathrm{Id}.$$

$T_2^{\otimes}$ and $V_{rs}^{\otimes}$, $r,s = 1,2$, are defined similarly. It is well known that $T_r^{\otimes}$, $r = 1,2$, are self-adjoint operators in H with

$$\mathcal{D} \subset \mathcal{D}(T_1^{\otimes}) \cap \mathcal{D}(T_2^{\otimes}), \tag{3.6}$$

and V_{rs}, $r,s = 1,2$, are bounded self-adjoint operators defined on all of $H^{\otimes}$, satisfying properties identical to those in (3.1) and (3.2).

Thus, form (3.4) we obtain a system of linear operators in $H^{\otimes}$:

$$W_r^{\otimes}(\underset{\sim}{\lambda}) := T_r^{\otimes} - \lambda_1 V_{r1}^{\otimes} - \lambda_2 V_{r2}^{\otimes} : H^{\otimes} \supset \mathcal{D}(T_r^{\otimes}) \to H^{\otimes}, \quad r = 1,2. \tag{3.7}$$

From now on, we omit the superscript $\otimes$ from induced operators: the space in which operators are acting will be clear from the context.

To write (3.7) more conveniently, we go one step further and consider the direct sum of two copies of $H^{\otimes}$:

$$H := H^{\otimes} \oplus H^{\otimes}$$

equipped with the usual inner product and norm: for $f = f_1 \oplus f_2$, $g = g_1 \oplus g_2 \in H$

$$(f,g) := (f_1,g_1)^{\otimes} + (f_2,g_2)^{\otimes};$$

$$\|\cdot\| := \sqrt{(.,.)} \quad .$$

We define a direct sum $T = T_1 \oplus T_2$ of the operators $T_r : H^{\otimes} \supset \mathcal{D}(T_r) \to H^{\otimes}$ by

$$\mathcal{D}(T) := \mathcal{D}(T_1) \oplus \mathcal{D}(T_2) \subset H,$$

$$T(f_1 \oplus f_2) := T_1 f_1 \oplus T_2 f_2.$$

T can be considered as a column vector of operators

$$T = \begin{bmatrix} T_1 \\ \\ T_2 \end{bmatrix}$$

and is a self-adjoint operator in H .

We write V for the matrix of operators

$$\begin{bmatrix} V_{11} & V_{12} \\ V_{21} & V_{22} \end{bmatrix} : H \to H$$

i.e. $V(f_1 \oplus f_2) = (V_{11}f_1 + V_{12}f_2) \oplus (V_{21}f_1 + V_{22}f_2)$

We denote by Λ the direct sum of multiplication operators

$$\Lambda := \lambda_1 \oplus \lambda_2 = \begin{bmatrix} \lambda_1 & \\ & \lambda_2 \end{bmatrix} : H \to H$$

$$\Lambda(f_1 \oplus f_2) := \lambda_1 f_1 \oplus \lambda_2 f_2$$

If composition of matrices of operators is denoted by "$\circ$", then (3.7) can be written

$$T - V \circ \Lambda : H \supset \mathcal{D}(T) \to H \tag{3.8}$$

The most important observation in discussing (3.8) is that the concept of co-factors can be formally carried over to the matrix V - note that the operators in different rows of V commute on $H^{\otimes}$. Thus writing

$$\hat{V} = \begin{bmatrix} V_{22} & -V_{12} \\ -V_{21} & V_{11} \end{bmatrix}$$

we obtain

$$\hat{V} \circ V = \begin{bmatrix} \Delta_o & 0 \\ 0 & \Delta_o \end{bmatrix} = \Delta_o \oplus \Delta_o : H \to H$$

where $\Delta_o = V_{11}V_{22} - V_{21}V_{12} : H^{\otimes} \to H^{\otimes}$.

(3.1) ensures that $\Delta_o : H^{\otimes} \to H^{\otimes}$ is a positive definite operator so that $\Delta_o \oplus \Delta_o : H \to H$ is also positive definite: for convenience we shall denote this latter operator simply by Δ_o; this should not lead to confusion.

Then, by Lemma 2.2

$$[.,.]^{\otimes} := (.,\Delta_o.)^{\otimes}, \quad |||\cdot|||^{\otimes} := \sqrt{[.,.]^{\otimes}}$$

define an inner product and norm on $H^{\otimes}$. We define

$$\mathcal{H}^{\otimes} := \overline{\{H^{\otimes}, [.,.]^{\otimes}\}}$$

and

$$\mathcal{H} := \mathcal{H}^{\otimes} \oplus \mathcal{H}^{\otimes}$$

Equivalently,

$$\mathcal{H} = \overline{\{H,[.,.]\}}$$

where, for $f = f_1 \oplus f_2$, $g = g_1 \oplus g_2 \in H$,

$$[f,g] := [f_1,g_1]^{\otimes} + [f_2,g_2]^{\otimes}, \quad |||\cdot||| := \sqrt{[.,.]} \ .$$

By Lemma 2.2, $\Delta_0 : H \to H$ has a positive definite extension to $\mathcal{H}$ - we use the same symbol to denote this extension - and $\Delta_0 \mathcal{H} \subset H$, so that

$$\Delta_0^{-1} : \mathcal{H} \supset \mathcal{D}(\Delta_0^{-1}) \to \mathcal{H}$$

is a densely defined, possibly unbounded, positive definite operator whose range is all of $\mathcal{H}$.

It is convenient at this point to introduce two other completions of $H^{\otimes}$:

$$\begin{aligned} \mathcal{H}_1^{\otimes} &:= \overline{\{H^{\otimes},[.,.]_1^{\otimes} := (.,V_{11}V_{22}.)^{\otimes}\}}, \quad |||\cdot|||_1^{\otimes} := \sqrt{[.,.]_1^{\otimes}} \ ; \\ \mathcal{H}_2^{\otimes} &:= \overline{\{H^{\otimes},[.,.]_2^{\otimes} := (.,-V_{21}V_{12}.)^{\otimes}\}}, \quad |||\cdot|||_2^{\otimes} := \sqrt{[.,.]_2^{\otimes}} \ . \end{aligned} \tag{3.9}$$

Now, premultiplying (3.8) by $\hat{V}$ and denoting $\hat{V} \circ T$ by Δ, we obtain the operator

$$\Delta - \Delta_0 \circ \Lambda \tag{3.10}$$

which can be considered as an operator in H or $\mathcal{H}$. Thus $\Delta = \Delta_1 \oplus \Delta_2$ where

$$\Delta_1 = V_{22}T_1 - V_{12}T_2$$

$$\Delta_2 = -V_{21}T_1 + V_{11}T_2 \ .$$

Finally, premultiplying (3.10) by Δ_0^{-1} we obtain

$$\Gamma - \Lambda \tag{3.11}$$

where $\Gamma = \Delta_0^{-1}\Delta$. We shall consider Γ to be an operator in H defined on the domain

$$\mathcal{D}(\Gamma) := \{f \in H | f \in \mathcal{D} \oplus \mathcal{D},\ \Delta f \in \mathcal{D}(\Delta_0^{-1})\}. \tag{3.12}$$

The first step will be to prove that Γ so defined is essentially self-adjoint. However, at this stage, it is not even clear that $\mathcal{D}(\Gamma)$ is dense in H. This is the subject of the next section. We conclude this section by introducing various mappings from tensor product spaces into the factor spaces. These mappings, referred to as factorisations, are defined as follows:

Fix $g_r \in H_{rs}$, (see (3.3)).

Given $f = \sum_{p=1}^{n} f_1^p \otimes f_2^p \in H_a^{\otimes}$, define the linear mapping

$$f \to [g_r, f]_{rs}$$

by

$$[g_1, f]_{1s} := \sum_{p=1}^{n} [g_1, f_1^p]_{1s} f_2^p \in H_2, \quad s = 1,2, \tag{3.13}$$

$$[g_2, f]_{2s} := \sum_{p=1}^{n} [g_2, f_2^p]_{2s} f_1^p \in H_1, \quad s = 1,2. \tag{3.14}$$

<u>Lemma 3.1:</u> For $f = \sum_{p=1}^{n} f_1^p \otimes f_2^p \in H_a^{\otimes}$

$$|||[g_1, f]_{11}|||_{22} \leq |||g_1|||_{11} |||f|||_1^{\otimes}$$
$$|||[g_1, f]_{12}|||_{21} \leq |||g_1|||_{12} |||f|||_2^{\otimes}$$
$$|||[g_2, f]_{21}|||_{12} \leq |||g_2|||_{21} |||f|||_2^{\otimes}$$
$$|||[g_2, f]_{22}|||_{11} \leq |||g_2|||_{22} |||f|||_1^{\otimes}$$

so that the mapping defined by (3.13) can be extended to bounded linear operators from $H_1^{\otimes}$ to H_{22} and $H_2^{\otimes}$ to H_{21} respectively, and those defined by (3.14) can be extended to bounded linear operators from $H_2^{\otimes}$ to H_{12} and $H_1^{\otimes}$ to H_{11}, respectively.

Proof: Given $f = \sum_{p=1}^{n} f_1^p \otimes f_2^p \in H_a^{\otimes}$, consider the finite dimensional subspaces

$$M_r := Sp[f_r^p,\ p = 1,\dots,n] \subset H_r$$

$$\dim M_r = m_r \leq n, \quad r = 1,2.$$

By Lemma 2.13, there exist complete totally orthogonal sets $\{\varphi_r^p\}_{p=1}^{m_r} \subset M_r$, $r = 1,2$. Then, for some coefficients a_{pq} $p = 1,\dots,m_1$, $q = 1,\dots,m_2$,

$$f = \sum_{p=1}^{m_1} \sum_{q=1}^{m_2} a_{pq}\varphi_1^p \otimes \varphi_2^q$$

Then

$$\begin{aligned}
|||[g_1,f]_{11}|||_{22}^2 &= |||\sum_p \sum_q a_{pq}[g_1,\varphi_1^p]_{11}\varphi_2^q|||_{22}^2 \\
&= \sum_q |\sum_p a_{pq}[g_1,\varphi_1^p]_{11}|^2 |||\varphi_2^q|||_{22}^2 \\
&= \sum_q |[g_1,\sum_p a_{pq}\varphi_1^p]_{11}|^2 |||\varphi_2^q|||_{22}^2 \\
&\leq |||g_1|||_{11}^2 \sum_q |||\sum_p a_{pq}\varphi_1^p|||_{11}^2 |||\varphi_2^q|||_{22}^2 \\
&= |||g_1|||_{11}^2 \sum_p \sum_q |a_{pq}|^2 |||\varphi_1^p|||_{11}^2 |||\varphi_2^q|||_{22}^2 \\
&= (|||g|||_{11} |||f|||_1^{\otimes})^2
\end{aligned}$$

This proves the first inequality; the others are proved similarly. □

Lemma 3.2: For g_1 and g_2 in appropriate spaces

(i) $[g_1 \otimes g_2,f]_1^{\otimes} = [g_1,[g_2,f]_{22}]_{11} \quad \forall f \in H_1^{\otimes}$;

(ii) $[g_1 \otimes g_2, f]^{\otimes}_2 = [g_2,[g_1,f]_{12}]_{21} \quad \forall f \in H^{\otimes}_2$;

(iii) $[g_1,[g_2,f]_{22}]_{11} = [g_2,[g_1,f]_{11}]_{22} \quad \forall f \in H^{\otimes}$;

(iv) $[g_1,[g_2,f]_{21}]_{12} = [g_2,[g_1,f]_{12}]_{21} \quad \forall f \in H^{\otimes}$.

Proof: The equalities are clear for elements $f \in H^{\otimes}_a$ and extend by continuity to the appropriate spaces; for (iii) and (iv) note that $H^{\otimes} = H^{\otimes}_1 \cap H^{\otimes}_2$. □

4 The denseness of $\mathcal{D}(\Gamma)$ in $\mathcal{H}$

Let A^2 denote the cartesian product of a set A with itself.

Firstly, we define a sequilinear form on $(\mathcal{D} \oplus \mathcal{D})^2$:

$$\gamma : (\mathcal{D} \oplus \mathcal{D})^2 \to \mathbb{C}$$
$$\gamma(f,g) := (f,\Delta g). \tag{4.1}$$

Lemma 4.1: For all $f \in \mathcal{D} \oplus \mathcal{D}$, $g \in \mathcal{D}(\Gamma)$

$$\gamma(f,g) = [f,\Gamma g].$$

Proof: For all $f \in \mathcal{D} \oplus \mathcal{D}$, $g \in \mathcal{D}(\Gamma)$

$$[f,\Gamma g] = [f,\Delta_0^{-1}\Delta g], \text{ by definition of } \Gamma$$
$$= (f,\Delta g)$$
$$= \gamma(f,g). \qquad \square$$

Now γ is used to define a class of operators in $\mathcal{H}$, $\{\Gamma^{\alpha,\beta} | \alpha,\beta \in \mathbb{C}\}$, as follows:

$$\mathcal{D}(\Gamma^{\alpha,\beta}) := \mathcal{D}(V_{11}^{-1}(T_1-\alpha))\otimes_a \mathcal{D}(V_{21}^{-1}(T_2-\alpha))$$
$$\oplus \mathcal{D}(V_{12}^{-1}(T_2-\beta))\otimes_a \mathcal{D}(V_{22}^{-1}(T_2+\beta)), \tag{4.2}$$

and for $g \in \mathcal{D}(\Gamma^{\alpha,\beta})$, $\Gamma^{\alpha,\beta} g$ is defined by the relationship

$$\gamma(f,g) = [f,\Gamma^{\alpha,\beta}g] \quad \forall f \in \mathcal{D} \oplus \mathcal{D} \tag{4.3}$$

That (4.3) defines a unique element $\Gamma^{\alpha,\beta}g \in \mathcal{H}$ follows from:

Lemma 4.2: For each $g = (g_{11} \otimes g_{12}) \oplus (g_{21} \otimes g_{22}) \in \mathcal{D}(\Gamma^{\alpha,\beta})$, $\gamma(.,g)$ defines a bounded conjugate linear functional on $\mathcal{D} \oplus \mathcal{D} \in H$, i.e. there is a constant $c(g) \geq 0$ such that

$$|\gamma(f,g)| \leq c(g)|||f||| \quad \forall f \in \mathcal{D} \oplus \mathcal{D}. \tag{4.4}$$

Proof: $\gamma(f,g) = \gamma_1(f_1,g_1) + \gamma_2(f_2,g_2)$ where $f = f_1 \oplus f_2$, $g = g_1 \oplus g_2$ and $\gamma_s(f_s,g_s) = (f_s,\Delta_s g_s)^{\otimes}$, $s = 1,2$. Then (4.3) can be written

$$\gamma_1(f_1,g_1) + \gamma_2(f_2,g_2) = [f_1,\Gamma_1^{\alpha}g_1]^{\otimes} + [f_2,\Gamma_2^{\beta}g_2]^{\otimes}.$$

We consider γ_1 defining Γ_1^{α}; γ_2 and Γ_2^{β} can be treated similarly.

For $g_1 = g_{11} \otimes g_{12} \in \mathcal{D}(\Gamma_1^{\alpha}) = \mathcal{D}(V_{11}^{-1}(T_1-\alpha)) \otimes_a \mathcal{D}(V_{21}^{-1}(T_2-\alpha))$, and

$$f_1 = \sum_{p=1}^{n} f_{11}^p \otimes f_{12}^p \in \mathcal{D},$$

$$\begin{aligned}
\gamma_1(f_1,g_1) &= (f_1,\Delta_1 g_1)^{\otimes} \\
&= \sum_{p=1}^{n} (f_{11}^p \otimes f_{12}^p,(V_{22}T_1-V_{12}T_2)(g_{11} \otimes g_{12}))^{\otimes} \\
&= \sum_{p=1}^{n} \{(f_{11}^p,T_1g_{11})_1(f_{12}^p,V_{22}g_{12})_2 - (f_{11}^p,V_{12}g_{11})_1(f_{12}^p,T_2g_{12})_2\} \\
&= \sum_{p=1}^{n} \{(f_{11}^p,T_1g_{11})_1[f_{12}^p,g_{12}]_{22} - [f_{11}^p,g_{11}]_{12}(f_{12}^p,T_2g_{12})_2\} \\
&= \sum_{p=1}^{n} \{(f_{11}^p,(T_1-\alpha V_{12})g_{11})_1[f_{12}^p,g_{12}]_{22} \\
&\qquad - [f_{11}^p,g_{11}]_{12}(f_{12}^p,(T_2-\alpha V_{22})g_{12})_2\} \\
&= (\sum_{p=1}^{n} [g_{12},f_{12}^p]_{22}f_{11}^p,(T_1-\alpha V_{12})g_{11})_1 \\
&\qquad - (\sum_{p=1}^{n} [g_{11},f_{11}^p]_{12}f_{12}^p,(T_2-\alpha V_{22})g_{12})_2 \\
&= [[g_{12},f_1]_{22},V_{11}^{-1}(T_1-\alpha V_{12})g_{11}]_{11} \\
&\qquad + [[g_{11},f_1]_{12}, V_{21}^{-1}(T_2-\alpha V_{22})g_{12}]_{21}.
\end{aligned}$$

Then, using the Cauchy inequality and Lemma 3.1,

$$|\gamma_1(f_1,g_1)| \leq |||[g_{12},f_1]_{22}|||_{11}|||V_{11}^{-1}(T_1-\alpha V_{12})g_{11}|||_{11}$$

$$+ |||[g_{11},f_1]_{12}|||_{21}|||V_{21}^{-1}(T_2-\alpha V_{22})g_{12}|||_{21}$$

$$\leq |||g_{12}|||_{22}|||f_1|||_1^{\otimes}|||V_{11}^{-1}(T_1-\alpha V_{12})g_{11}|||_{11}$$

$$+ |||g_{11}|||_{12}|||f_1|||_2^{\otimes}|||V_{21}^{-1}(T_2-\alpha V_{22})g_{12}|||_{21}$$

$$\leq (|||g_{12}|||_{22}|||V_{11}^{-1}(T_1-\alpha V_{12})g_{11}|||_{11}$$

$$+ |||g_{11}|||_{12}|||V_{21}^{-1}(T_2-\alpha V_{22})g_{12}|||_{21})|||f_1|||^{\otimes}$$

since $|||f_1|||_s \leq |||f_1|||$, $s = 1,2$. Thus

$$|\gamma_1(f_1,g_1)| \leq c(g_1)|||f_1|||^{\otimes}$$

where $c(g_1)$ is a constant for fixed g_1.

Similarly, for $g_2 = g_{21} \otimes g_{22} \in \mathcal{D}(\Gamma_2^{\beta})$ and $f_2 \in \mathcal{D}$

$$|\gamma_2(f_2,g_2)| \leq c(g_2)|||f_2|||^{\otimes} \quad .$$

Thus, for $g = (g_{11} \otimes g_{12}) \oplus (g_{21} \otimes g_{22}) \in \mathcal{D}(\Gamma^{\alpha,\beta})$

$$|\gamma(f,g)| \leq |\gamma_1(f_1,g_1)| + |\gamma_2(f_2,g_2)|$$

$$\leq c(g)|||f||| \quad \forall f \in \mathcal{D} \oplus \mathcal{D},$$

and so $\gamma(.,g) : H \supset \mathcal{D} + \mathcal{D} \to \mathbb{C}$ is a bounded conjugate linear functional on $\mathcal{D} \oplus \mathcal{D}$ which is dense in H. □

By the Riesz representation theorem, there exists a unique element $\varphi(g) = \varphi_1(g) \oplus \varphi_2(g) \in H$ such that

$$\gamma(f,g) = [f,\varphi(g)] \quad \forall f \in \mathcal{D} \oplus \mathcal{D}$$

Then (4.3) defines $\Gamma^{\alpha,\beta}g := \varphi(g)$, and the definition of $\Gamma^{\alpha,\beta}$ can be extended to all of $\mathcal{D}(\Gamma^{\alpha,\beta})$ by linearity.

Definition 4.3: Given any set $S \subset \mathbb{C} \times \mathbb{C}$, the operators $\Gamma^S : H \supset \mathcal{D}(\Gamma^S) \to H$ is defined as follows:

$$\mathcal{D}(\Gamma^S) := \sum_{(\alpha,\beta)\in S} \mathcal{D}(\Gamma^{\alpha,\beta}) \ ;$$

for $g = \sum_{i=1}^{n} g_i \in \mathcal{D}(\Gamma^S)$ where $g_i \in \mathcal{D}(\Gamma^{\alpha_i,\beta_i})$, $(\alpha_i,\beta_i) \in S$, $i = 1,\ldots,n < \infty$:

$$\Gamma^S g := \sum_{i=1}^{n} \Gamma^{\alpha_i,\beta_i} g_i \ .$$

This definition makes sense since, if $(\alpha_1,\beta_1) \in S$ and $(\alpha_2,\beta_2) \in S$ and $g \in \bigcap_{i=1}^{2} \mathcal{D}(\Gamma^{\alpha_i,\beta_i})$, then it is clear from (4.3) that $\Gamma^{\alpha_1,\beta_1}g = \Gamma^{\alpha_2,\beta_2}g$.

Lemma 4.4: For any set $S \subset \mathbb{C} \times \mathbb{C}$, $\Gamma^S \subset \Gamma$.

Proof: For any $(\alpha,\beta) \in \mathbb{C} \times \mathbb{C}$ and for all $f \in \mathcal{D} \oplus \mathcal{D}$, $g \in \mathcal{D}(\Gamma^{\alpha,\beta})$

$$[f,\Gamma^{\alpha,\beta}g] = \gamma(f,g) = (f,\Delta g). \qquad (4.5)$$

Now, $\mathcal{D} \oplus \mathcal{D}$ is dense in H, and since

$$|||f|||^2 = (f,\Delta_0 f) \leq \|\Delta_0\| \ \|f\|^2 \qquad \forall f \in \mathcal{D} \oplus \mathcal{D},$$

(4.4) shows that $\gamma(.,g)$ is a bounded conjugate linear functional on $\mathcal{D} \oplus \mathcal{D} \subset H$ so that (4.5) can be extended by continuity to all $f \in H$. However, $\mathcal{D}(\Delta_0^{-1}) \subset H$ (see Lemma 2.2), so that, for all $f \in \mathcal{D}(\Delta_0^{-1})$

$$[f,\Gamma^{\alpha,\beta}g] = (\Delta_0\Delta_0^{-1}f,\Delta g) = [\Delta_0^{-1}f,\Delta g] \qquad (4.6)$$

But Δ_0^{-1} is a self-adjoint operator in H since Δ_0 is self-adjoint, so that it follows from (4.6) that

$$\Delta g \in \mathcal{D}(\Delta_0^{-1}) \quad \Delta_0^{-1}\Delta g = \Gamma^{\alpha,\beta}g.$$

Comparing this with (3.12) we see that

$$g \in \mathcal{D}(\Gamma), \quad \Gamma g = \Gamma^{\alpha,\beta}g.$$

Therefore $\Gamma^{\alpha,\beta} \subset \Gamma$ for any $(\alpha,\beta) \in \mathbb{C} \times \mathbb{C}$, so that for any $S \subset \mathbb{C} \times \mathbb{C}$, $\Gamma^S \subset \Gamma$.

□

Corollary 4.5: Γ is densely defined in $\mathcal{H}$.

Proof: Choose $\alpha \in \rho(T_1) \cap \rho(T_2)$, $\beta \in \rho(T_1) \cap \rho(-T_2)$; any non-real α and β will suffice since T_1 and T_2 are self-adjoint. Then Lemma 2.3(ii) shows that each factor space appearing in (4.2) is dense in the appropriate space H_1 or H_2, so that $\mathcal{D}(\Gamma^{\alpha,\beta})$ is dense in H and so also dense in $\mathcal{H}$. Since $\Gamma^{\alpha,\beta} \subset \Gamma$, the result follows. □

5 Essential self-adjointness

At this stage we know that Γ is a densely defined symmetric operator in H. In order to ensure that Γ is essentially self-adjoint, some further assumptions are required.

Condition 5.1: There exist sets of disjoint open intervals $\{I_\nu\}$ and $\{J_\mu\}$, each of which covers the real line except for an at most countable set of points N_I and N_J, respectively, such that for each pair (ν,μ), for all $(\alpha,\beta) \in I_\nu \times I_\mu$

$$\text{at least one component of } T-V\circ\binom{i}{\alpha} : H \supset \mathcal{D}(T) \to H \text{ has a bounded inverse on } H^{\otimes} \tag{5.1}$$

and

$$\text{at least one component of } T-V\circ\binom{\beta}{i} : H \supset \mathcal{D}(T) \to H \text{ has a bounded inverse on } H^{\otimes}. \tag{5.2}$$

For convenience, we have stated these assumptions in terms of operators in H and $H^{\otimes}$. However, if we define

$$R_r(\lambda_1,\lambda_2) := [W_r(\lambda_1,\lambda_2)]^{-1} : H_r \to H_r, \quad r = 1,2 \tag{5.3}$$

$$\left.\begin{array}{ll}
\text{then (5.1) is equivalent to} & \\
\quad R_1(i,\alpha) \in B(H_1) & (11) \\
\text{or} & \\
\quad R_2(i,\alpha) \in B(H_2) & (21) \\
\text{and (5.2) is equivalent to} & \\
\quad R_1(\beta,i) \in B(H_1) & (12) \\
\text{or} & \\
\quad R_2(\beta,i) \in B(H_2) & (22)
\end{array}\right\} \tag{5.4}$$

These follow easily from the definitions of induced operators.

Then, (5.1) is sufficient to show that Γ_1 is essentially self-adjoint and (5.2) is sufficient to show that Γ_2 is essentially self-adjoint. We shall prove the first of these claims; the latter follows analogously.

We define functions of a complex variable z: for $g_1 \in H_{11}$, $g_2 \in H_{21}$ and $f \in H^{\otimes}$

$$\begin{aligned}
\chi_1(z) &:= [[g_1,f]_{11},R_2(i,z)V_{21}g_2]_{22} \\
&= (V_{22}[g_1,f]_{11},R_2(i,z)V_{21}g_2)_2 \\
\chi_2(z) &:= [[g_2,f]_{21},R_1(i,z)V_{11}g_1]_{12} \\
&= (V_{12}[g_2,f]_{21},R_1(i,z)V_{11}g_1)_1
\end{aligned} \tag{5.5}$$

Lemma 2.7 shows that $R_2(i,z) \in B(H_2)$ for Im $z < 0$ and that $R_1(i,z) \in B(H_1)$ for Im > 0. Then, Lemma 2.11 shows that χ_1 is analytic in $\mathbb{C}^-$ and χ_2 is analytic in $\mathbb{C}^+$ where $\mathbb{C}^{\pm} := \{z \in \mathbb{C} | \text{Im } z \gtrless 0\}$.

Now, Condition 5.1 implies that, for each ν, either χ_1 or χ_2 can be defined for $z = \alpha$ for all $\alpha \in I_\nu$, so that, by Lemma 2.11, either χ_1 or χ_2 can be defined in some open set $G_\nu \subset \mathbb{C}$ containing I_ν. Let $G_\nu^{\pm} := G_\nu \cap \mathbb{C}^{\pm}$. Then, for each ν, both χ_1 and χ_2 are defined and analytic in at least one of G_ν^+ and G_ν^-. Denote the appropriate set by G_ν^0. Let $S = \bigcup_\nu G_\nu^0$, and consider the operator $\Gamma_1^S \subset \Gamma_1$; we show that Γ_1^S has deficiency indices (0,0), so that Γ_1^S is essentially self-adjoint. Let $f \in H^{\otimes}$ be such that

$$[f,(\Gamma_1^S-i)g]^{\otimes} = 0 \quad \forall g \in \mathcal{D}(\Gamma_1^S).$$

Then, for any $\alpha \in S$ and $g = g_1 \otimes g_2 \in \mathcal{D}(\Gamma_1^\alpha)$

$$\begin{aligned}
0 &= [f,(\Gamma_1^\alpha-i)g]^{\otimes} \\
&= [f,\Gamma_1^\alpha g]^{\otimes} - [f,ig]^{\otimes} \\
&= [[g_2,f]_{22},V_{11}^{-1}(T-\alpha V_{12})g_1]_{11} + [[g_1,f]_{12},V_{21}^{-1}(T_2-\alpha V_{22})g_2]_{21} \\
&\qquad - [f,ig_1 \otimes g_2]_1^{\otimes} - [f,ig_1 \otimes g_2]_2^{\otimes}
\end{aligned}$$

(since the calculation in the proof of Lemma 4.2 extends to all $f \in H^{\otimes}$, and $[.,.]^{\otimes} = [.,.]_1^{\otimes} + [.,.]_2^{\otimes}$)

$$= [[g_2,f]_{22},(V_{11}^{-1}(T-\alpha V_{12})-i)g_1]_{11} + [[g_1,f]_{12},(V_{21}^{-1}(T_2-\alpha V_{22})-i)g_2]_{21}$$

(using Lemma 3.2).

Let

$$h_1 := (V_{11}^{-1}(T_1-\alpha V_{12})-i)g_1 \text{ so that } g_1 = R_1(i,\alpha)V_{11}h_1$$

and

$$h_2 := (V_{21}^{-1}(T_2-\alpha V_{22})-i)g_2 \text{ so that } g_2 = R_2(i,\alpha)V_{21}h_2$$

By Lemma 2.3, $R(V_{r1}^{-1}(T_r-iV_r-\alpha V_{r2}) = H_{r1}$, $r = 1,2$, so that, for all $h_1 \in H_{11}$, $h_2 \in H_{21}$

$$0 = [[R_2(i,\alpha)V_{21}h_2,f]_{22},h_1]_{11} + [[R_1(i,\alpha)V_{11}h_1,f]_{12},h_2]_{21}.$$

Using Lemma 3.2, we obtain

$$[[h_1,f]_{11},R_2(i,\alpha)V_{21}h_2]_{22} = -[[h_2,f]_{21},R_1(i,\alpha)V_{11}h_1]_{12}$$

i.e. $\chi_1(\alpha) = -\chi_2(\alpha) \quad \forall\alpha \in S$.

Standard theory of analytic continuation shows that the function

$$\chi(z) := \begin{cases} \chi_1(z) & \text{on } \mathbb{C}^- \cup S \\ -\chi_2(z) & \text{on } \mathbb{C}^+ \cup S \end{cases}$$

is analytic in the whole complex plane except for the exceptional set $N_I = \mathbb{R} \setminus \bigcup_\nu I_\nu$.

Now, from the definition of χ_1 and Lemma 2.6(iii) and Lemma 3.1 it follows that, for any $f \in H^{\otimes}$, $g_1 \in H_{11}$, $g_2 \in H_{21}$,

$$|\chi_1(z)| = |[[g_1,f]_{11},R_2(i,z)V_{21}g_2]_{22}|$$

$$\leq |||[g_1,f]_{11}|||_{22}|||R_2(i,z)V_{21}g_2|||_{22}$$

$$\leq |||g_1|||_{11}|||f|||_1^{\otimes} \frac{1}{\sqrt{2(-1)\operatorname{Im} z}} |||g_2|||_{21} \quad \text{for} \quad \operatorname{Im} z < 0.$$

Similarly

$$|\chi_2(z)| \leq |||g_2|||_{21}|||f|||_2^{\otimes} \frac{1}{\sqrt{2.1.\operatorname{Im} z}} |||g_1|||_{11} \quad \text{for} \quad \operatorname{Im} z > 0.$$

Thus, for any $g_1 \in H_{11}$, $g_2 \in H_{21}$, there exist a constant $c > 0$ such that, for all $z \in \mathbb{C}\setminus N_I$

$$|\sqrt{(\operatorname{Im} z)}\chi(z)| < c \tag{5.6}$$

Now, consider an isolated point in the exceptional set N_I; without loss of generality assume the point is $z = 0$. Then there exists a deleted neighborhood $\{z \in \mathbb{C}|0 < |z| < \delta\}$ within which $\chi(z)$ is analytic. Consider a circular contour C: $|z| = r$, $0 < r < \delta$:

$$\left|\int_C \chi(z)z^k dz\right| = \left|\int_0^{2\pi} \chi(re^{i\vartheta})r^k e^{ik\vartheta} ire^{i\vartheta}d\vartheta\right|$$

$$\leq cr^{k+1/2}\int_0^{2\pi} \frac{1}{\sqrt{|\sin\vartheta|}}\, d\vartheta \quad \text{using (5.6)}$$

$$= 4cr^{k+1/2}\int_0^{\pi/2} (\sin\vartheta)^{-1/2}d\vartheta$$

$$\leq 4cr^{k+1/2}\int_0^{\pi/2} \sqrt{\frac{\pi}{2}}\,\vartheta^{-1/2}d\vartheta = c'r^{k+1/2} \to 0$$

as $r \to 0$ for all $k \geq 0$. Thus, in the Laurent expansion of $\chi(z)$ about the singularity at $z = 0$ the coefficient of z^k is 0 for $k \leq 0$.

Thus the isolated singularity is removeable, and moreover, $\chi(z)$ takes the value 0 at this point.

A general functional analytical argument shows that all points in N_I are removeable singularities of χ.

Assume otherwise: then, having removed all the isolated points of N_I, we are left with a set N_I^0 containing no isolated points. $\overline{N_I^0} \subset \overline{N_I} = N_I$ since N_I is closed and so $\overline{N_I^0}$ is countable. Also $\overline{N_I^0}$ is a complete metric space. Therefore

$$\overline{N_I^0} = \bigcup_{x \in \overline{N_I^0}} \{x\}$$

and for each $x \in \overline{N_I^0}$ $\{x\}$ is nowhere dense in $\overline{N_I^0}$. Thus, $\overline{N_I^0}$ is of the first category (a countable union of nowhere dense sets) and the Baire Theorem shows that $\overline{N_I^0}$ must be empty.

Thus, $\chi(z)$ is an entire function. Consider the Taylor series expansion of $\chi(z)$ about $z = 0$.

The above calculation shows that

$$\left|\int_C \chi(z) z^k dz\right| \leq c' r^{k+1/2} \begin{cases} \to 0 \text{ as } r \to \infty \text{ for } k \leq -1 \\ \to 0 \text{ as } r \to 0 \text{ for } k = 0 \end{cases}$$

i.e. all coefficients in the Taylor series vanish:

$$\chi(z) \equiv 0.$$

Now consider (5.5). For any $g_1 \in H_{11}$, $g_2 \in H_{21}$

$$0 = \chi_1(z) = [[g_1, f]_{11}, R_2(i,z)V_{21}g_2]_{22}$$

i.e. $0 = [g_1 \otimes R_2(i,z)V_{21}g_2, f]_1^{\otimes}$ (using Lemma 3.2). (5.7)

Now, $R_2(i,z)V_{21}H_{21} = \mathcal{D}(V_{21}^{-1}(T_2 - z))$ which is dense in H_2, so that (5.7) says that

$$[g,f]_1^{\otimes} = 0 \quad \forall g \text{ in a dense subspace of } H_1^{\otimes}$$

$$\Rightarrow f = 0 \in H_1^{\otimes}.$$

But the zero elements in $H^{\otimes}$, $H_1^{\otimes}$, $H_2^{\otimes}$ and $H^{\otimes}$ can be identified so that

$$f = 0 \in H^{\otimes},$$

and the negative deficiency index of Γ_1^S is zero.

Considering Condition 5.1 and noting that

$$[T-V\circ(\tfrac{i}{\alpha})]^* = T-V\circ(\tfrac{-i}{\alpha})$$

we see that (5.1) will hold with -i in place of i.

An identical argument will now show that, for some $S \subset \mathbb{C}$,

$$[f,(\Gamma_1^S + i)g]^{\otimes} = 0 \quad \forall g \in \mathcal{D}(\Gamma_1^S)$$

$$\Rightarrow f = 0$$

so that the positive deficiency index of Γ_1^S is also zero, and Γ_1^S is essentially self-adjoint. Since Γ_1 is a symmetric extension of Γ_1^S, Γ_1 is also essentially self-adjoint.

Finally, in an identical fashion (5.2) will allow us to prove that Γ_2 is essentially self-adjoint, and so we have:

Theorem 5.2: Let Condition 5.1 hold. Then

$$\Gamma = \begin{bmatrix} \Gamma_1 \\ \Gamma_2 \end{bmatrix} : H \supset \mathcal{D}(\Gamma) \to H$$

is essentially self-adjoint.

6 Finite dimensional approximations

We define finite dimensional subspaces H^n of H. These can also be considered as subspaces of $\mathcal{H}$.

Let $\{f_k \oplus g_k\}_{k=1}^{\infty}$ be a dense set in $\mathcal{R}(\Gamma - i)$ which itself is dense in $\mathcal{H}$ since Γ is essentially self-adjoint, and let

$$\Phi_k := \varphi_k \oplus \psi_k := (\Gamma - i)^{-1}(f_k \oplus g_k) \in \mathcal{D}(\Gamma) \subset \mathcal{D} \oplus \mathcal{D} \subset \mathcal{D}(T) \qquad (6.1)$$

$$k = 1,2,\ldots$$

Thus, for all k,

$$\varphi_k = \sum_{p=1}^{M_k} \varphi_{k1}^p \otimes \varphi_{k2}^p, \; \varphi_{kr}^p \in \mathcal{D}(T_r), \; r = 1,2, \quad (M_k < \infty)$$

$$\psi_k = \sum_{q=1}^{N_k} \psi_{k1}^q \otimes \psi_{k2}^q, \; \psi_{kr}^q \in \mathcal{D}(T_r), \; r = 1,2, \quad (N_k < \infty).$$

Further we choose a set of elements $\{w_{kr}\}_{k=1}^{\infty}$ which is dense in H_r, $r = 1,2$, and let

$$\omega_{kr} = (T_r - i)^{-1} w_{kr} \subset \mathcal{D}(T_r) \quad k = 1,2,\ldots, \quad r = 1,2. \qquad (6.2)$$

We define finite dimensional subspaces H_r^n of H_r, $r = 1,2$, by

$$H_r^n := \mathrm{Sp}\Big[\bigcup_{k=1}^{n} \{\varphi_{kr}^p,\ p=1,\ldots,M_k;\ \psi_{kr}^q,\ q=1,\ldots,N_k;\ \omega_{kr}\}\Big], \quad r = 1,2.$$

Then we define a subspace H^n of H by

$$H^n := (H_1^n \otimes H_2^n) \oplus (H_1^n \otimes H_2^n).$$

Clearly, H^n is finite dimensional and $H^i \subseteq H^j$ for $i < j$.

In H^n we define a linear operator Γ^n using the sequilinear form (4.1):

$$[u,\Gamma^n v] := \gamma(u,v) \quad \forall u,v \in H^n.$$

It is clear that Γ^n is self-adjoint in $\{H^n,[.,.]\}$.

Let $P^n : \mathcal{H} \to H^n$ be the orthogonal projection of $\mathcal{H}$ on H^n. By definition, Γ^n is the restriction of Γ to H^n (see Lemma 4.1) and can be regarded as an operator on $\mathcal{H}$ given by

$$\Gamma^n := P^n \Gamma P^n : \mathcal{H} \to H^n \subset \mathcal{H}.$$

On the other hand, let $Q^n : H \to H^n$ be the orthogonal projection of H on H^n. It is easily verified that

$$Q^n = (Q_1^n \otimes Q_2^n) \oplus (Q_1^n \otimes Q_2^n) \tag{6.3}$$

where Q_r^n is the orthogonal projection of H_r on H_r^n, $r = 1,2$. Let

$$T^n := Q^n T Q^n$$

$$V^n := Q^n V Q^n \tag{6.4}$$

$$\Delta_0^n := \det V^n$$

$$\Delta^n := \hat{V}^n \circ T^n, \text{ where } \hat{} \text{ denotes the cofactor matrix.}$$

Careful consideration of these operators shows that they can be constructed from the operators

$$T_r^n := Q_r^n T_r Q_r^n : H_r \to H_r, \quad r = 1,2,$$

$$V_{rs}^n := Q_r^n V_{rs} Q_r^n, \; s = 1,2 : H_r \to H_r, \quad r = 1,2 \tag{6.5}$$

in the same way as T,V,Δ_0 and Δ are constructed from T_r, V_{rs}, $s = 1,2$, $r = 1,2$.

Now, regarding the operators in (6.5) and (6.4) as acting on H_r^n, $r = 1,2$, and H^n respectively, it is clear that $(\Delta_0^n)^{-1} : H^n \to H^n$ exists and we can define

$$G^n := (\Delta_0^n)^{-1}\Delta^n : H^n \to H^n.$$

Lemma 6.1: $G^n : H^n \to H^n$ is identical to $\Gamma^n : H^n \to H^n$.

Proof: For all $u,v \in H^n$

$$\begin{aligned}
[u,\Gamma^n v] &= \gamma(u,v) \\
&= (u,\Delta v) && \text{(see (4.1))} \\
&= (u,\Delta^n v) && \text{(since } u,v \in H^n\text{)} \\
&= (u,\Delta_0^n(\Delta_0^n)^{-1}\Delta^n v) \\
&= (u,\Delta_0(\Delta_0^n)^{-1}\Delta^n v) \\
&= [u,G^n v],
\end{aligned}$$

i.e. $\Gamma^n v = G^n v \quad \forall v \in H^n$. □

Lemma 6.2: $(\Gamma^n - i)^{-1}P^n f \overset{H}{\to} (\overline{\Gamma} - i)^{-1} f \quad \forall f \in H,$ (6.6)

where $\overline{\Gamma}$ denotes the closure of Γ.

Proof: Since Γ^n is self-adjoint on $\{H^n,[.,.]\}$

$$|||(\Gamma^n - i)^{-1} f||| \leq |||f||| \quad \forall f \in H^n$$

so that

$$|||(\Gamma^n - i)^{-1}P^n f||| \leq |||P^n f||| \leq |||f||| \quad \forall f \in H. \tag{6.7}$$

Now, for all Φ_k defined in (6.1), $\Phi_k \in H^n$ provided $n \geq k$ and

$$\Gamma^n\Phi_k = P^n\Gamma\Phi_k.$$

Therefore,

$$(\Gamma^n - i)\Phi_k = P^n(\Gamma - i)\Phi_k$$

$$\Rightarrow \Phi_k = (\Gamma^n - i)^{-1}P^n(\Gamma - i)\Phi_k \qquad \forall n \geq k. \tag{6.8}$$

Then, for all $f \in H$

$$|||(\Gamma^n - i)^{-1}P^n f \;-\; (\overline{\Gamma} - i)^{-1} f|||$$

$$= |||(\Gamma^n - i)^{-1}P^n(f - (\overline{\Gamma} - i)\Phi_k) - (\overline{\Gamma} - i)^{-1}(f - (\overline{\Gamma} - i)\Phi_k)||| \quad \text{(using (6.8))}$$

$$\leq 2|||f - (\overline{\Gamma} - i)\Phi_k||| \quad \text{(using (6.7))}.$$

Since $\{\Phi_k\}_{k=1}^{\infty}$ is such that $\{(\Gamma - i)\Phi_k\}_{k=1}^{\infty}$ is dense in H, Φ_k can be chosen to make this arbitrarily small, and the result is proved. □

The commutativity of the components of $\Gamma = \Gamma_1 \oplus \Gamma_2$ now follows. From standard results for finite dimensional multiparameter problems (see e.g. Sleeman [9]) the components of $G^n = G_1^n \oplus G_2^n$ commute, and since $G^n = \Gamma^n$, we have

$$\Gamma_1^n \Gamma_2^n = \Gamma_2^n \Gamma_1^n \quad \text{on} \quad H_1^n \otimes H_2^n.$$

Then, for all $f \in H^{\otimes}$

$$(\Gamma_1^n - i)^{-1}P^n(\Gamma_2^n - i)^{-1}P^n f = (\Gamma_2^n - i)^{-1}P^n(\Gamma_1^n - i)^{-1}P^n f. \tag{6.9}$$

Let us define

$$R_s^n := (\Gamma_s^n - i)^{-1}, \quad R_s := (\overline{\Gamma}_s - i)^{-1}, \quad s = 1,2.$$

Then, for all $f \in H^{\otimes}$

$$|||R_1^n P^n R_2^n P^n f - R_1 R_2 f|||^{\otimes}$$

$$\leq |||R_1^n P^n (R_2^n P^n - R_2) f|||^{\otimes} + |||(R_1^n P^n - R_1) R_2 f|||^{\otimes}$$

$$\leq |||(R_2^n P^n - R_2) f|||^{\otimes} + |||(R_1^n P^n - R_1) R_2 f|||^{\otimes} \quad \text{(by (6.7))}$$

$$\to 0 \quad \text{as } n \to \infty \quad \text{by (6.6)};$$

i.e. $R_1^n P^n R_2^n P^n f \xrightarrow{H^{\otimes}} R_1 R_2 f$ as $n \to \infty$.

Similarly,

$$R_2^n P^n R_1^n P^n f \xrightarrow{H^{\otimes}} R_2 R_1 f \quad \text{as} \quad n \to \infty,$$

so that, by letting $n \to \infty$ in (6.9),

$$R_1 R_2 f = R_2 R_1 f \quad \forall f \in H^{\otimes}. \tag{6.10}$$

By a standard result for self-adjoint operators ([3], p.247) (6.10) implies that $\overline{\Gamma}_1$ and $\overline{\Gamma}_2$ commute in the sense that their spectral families commute.

Finally in this section we derive an initial generalization to this two parameter problem of a standard identity for finite dimensional problems. For our finite dimensional problems involving the operators in (6.4) it is well known that

$$T^n f = V^n \circ \Gamma^n f \quad \forall f \in H^n \tag{6.11}$$

(See e.g. [9], p.32 where the result is stated for the bounded operator situation).

Let

$$\{P_{rs}(\lambda), \quad -\infty < \lambda < \infty\} \tag{6.12}$$

be the spectral family of $|V_{rs}| : H^{\otimes} \to H^{\otimes}$, $r,s = 1,2$. Then, we define

$$P_{rs,\varepsilon} := P_{rs}((\varepsilon, 1-\varepsilon)), \quad 0 < \varepsilon < \frac{1}{2} .$$

We note that, since $0 < |V_{rs}| < 1$, $r,s = 1,2$, it follows that

$$\forall f \in H^{\otimes}, \quad P_{rs,\varepsilon} f \to f \quad \text{as} \quad \varepsilon \to 0, \quad r,s = 1,2. \tag{6.13}$$

Lemma 6.3: $\|P_{rr,\varepsilon} u\|^{\otimes} \leq \sqrt{\frac{2}{\varepsilon}}\, |||u|||^{\otimes}$, $\forall u \in H^{\otimes}$, $r = 1,2$.

Proof: Making an obvious extension of the notation introduced in Definition 2.1, we have

$$\begin{aligned} 0 &\leq (V_{11} + V_{21})^2 \\ &= (V_{11} + V_{21})(V_{11} + V_{21}) \\ &= (V_{11} + V_{21})(-V_{12} + V_{22}) \\ &= \Delta_o - V_{11}V_{12} + V_{21}V_{22}. \end{aligned}$$

Thus, using the notation $A \leq B$ to mean $(Au,u) \leq (Bu,u)$ $\forall u$, we have

$$V_{11}V_{12} - V_{21}V_{22} \leq \Delta_o,$$

so that

$$V_{11}V_{12} \leq \Delta_o, \quad -V_{21}V_{22} \leq \Delta_o.$$

Then, for all $u \in H^{\otimes}$

$$\begin{aligned} &[P_{11,\varepsilon}u, P_{11,\varepsilon}u]^{\otimes} \\ &= (P_{11,\varepsilon}u, \Delta_o P_{11,\varepsilon}u)^{\otimes} \\ &\geq (P_{11,\varepsilon}u, V_{11}V_{12}P_{11,\varepsilon}u)^{\otimes} \\ &= \int_{(\varepsilon,1-\varepsilon)} \lambda(1-\lambda)\, d\|P_{11}(\lambda)u\|^{\otimes 2} \end{aligned}$$

$$\geq \varepsilon(1-\varepsilon)\|P_{11,\varepsilon}u\|^{\otimes 2}$$

$$> \frac{\varepsilon}{2}\|P_{11,\varepsilon}u\|^{\otimes 2} \qquad \text{for } 0 < \varepsilon < \tfrac{1}{2}.$$

Now, since $P_{11}(\lambda)$ commutes with Δ_o,

$$[(Id-P_{11,\varepsilon})u, P_{11,\varepsilon}u]^{\otimes} = ((Id-P_{11,\varepsilon})u, \Delta_o P_{11,\varepsilon}u)^{\otimes}$$

$$= ((Id-P_{11,\varepsilon})\Delta_o^{1/2}u, P_{11,\varepsilon}\Delta_o^{1/2}u)^{\otimes} = 0$$

so that

$$|||u|||^{\otimes 2} = |||(Id-P_{11,\varepsilon})u|||^{\otimes 2} + |||P_{11,\varepsilon}u|||^{\otimes 2}.$$

Thus

$$\|P_{11,\varepsilon}u\|^{\otimes} \leq \sqrt{\frac{2}{\varepsilon}}\,|||u|||^{\otimes}.$$

The other estimate is proved in an identical manner. □

We now define

$$P_\varepsilon : H \to H$$
$$P_\varepsilon := P_{22,\varepsilon} \oplus P_{11,\varepsilon} \tag{6.14}$$

Note the order of the terms in this definition of P_ε; it follows that P_ε commutes with T.

<u>Corollary 6.4:</u> $\|P_\varepsilon u\| \leq \sqrt{\frac{2}{\varepsilon}}\,|||u||| \quad \forall u \in H.$

<u>Proof:</u> $\|P_\varepsilon u\|^2 = \|P_{22,\varepsilon}u_1\|^{\otimes^2} + \|P_{11,\varepsilon}u_2\|^{\otimes^2}, \quad u = u_1 \oplus u_2$

$$\leq \frac{2}{\varepsilon}\,|||u_1|||^{\otimes^2} + \frac{2}{\varepsilon}\,|||u_2|||^{\otimes^2}$$

$$= \frac{2}{\varepsilon}\,|||u|||^2. \qquad \square$$

It follows that P_ε can be extended to a bounded linear operator from $\mathcal{H} \to H$; we assume whenever necessary that the extension has been made.

Now, for $g \in \mathcal{H}$, let

$$f := (\overline{\Gamma}-i)^{-1}g, \; f^n := (\Gamma^n-i)^{-1}P^n g \in H^n.$$

Then

$$\lim_{n\to\infty} f^n = f \qquad \text{(by (6.6))},$$

$$\lim_{n\to\infty} (\Gamma^n-i)f^n = g = (\overline{\Gamma}-i)f,$$

so that

$$\lim_{n\to\infty} \Gamma^n f^n = \overline{\Gamma}f. \tag{6.15}$$

From (6.11)

$$T^n f^n = V^n \circ \Gamma^n f^n$$

so that

$$P_\varepsilon T^n f^n = P_\varepsilon V^n \circ \Gamma^n f^n$$

and since P_ε and T^n commute, we can conclude that

$$T^n P_\varepsilon f^n = P_\varepsilon V^n \circ \Gamma^n f^n. \tag{6.16}$$

Let

$$\tilde{\omega}_{k\ell} = (\omega_{k1} \otimes \omega_{k2}) \oplus (\omega_{\ell 1} \otimes \omega_{\ell 2}), \qquad k,\ell = 1,2,\ldots$$

where ω_{kr} is defined in (6.2). Then for $n \geq k,\ell$, $\tilde{\omega}_{k\ell} \in H^n \subset \mathcal{D}(T)$.

Then, from (6.16), for $n \geq k,\ell$

$$(\tilde{\omega}_{k\ell}, T^n P_\varepsilon f^n) = (\tilde{\omega}_{k\ell}, P_\varepsilon V^n \circ \Gamma^n f^n)$$

$$\Rightarrow (Q^n T\tilde{\omega}_{k\ell}, P_\varepsilon f^n) = (\tilde{\omega}_{k\ell}, P_\varepsilon V^n \circ \Gamma^n f^n).$$

Letting $n \to \infty$, we obtain, using (6.15)

$$(T\tilde{\omega}_{k\ell}, P_\varepsilon f) = (\tilde{\omega}_{k\ell}, P_\varepsilon V \circ \overline{\Gamma} f) \quad \forall f \in \mathcal{D}(\overline{\Gamma}).$$

This holds for all $\tilde{\omega}_{k\ell}$ and since $\{\tilde{\omega}_{k\ell}\}_{k,\ell=1}^{\infty}$ is dense in $\mathcal{D}(T)$, it follows from the self-adjointness of T that

$$\begin{aligned} &P_\varepsilon f \in \mathcal{D}(T) \\ &TP_\varepsilon f = P_\varepsilon V \circ \overline{\Gamma} f \quad \forall f \in \mathcal{D}(\overline{\Gamma}). \end{aligned} \tag{6.18}$$

This is an initial generalisation of the finite dimensional result (6.11). Our next task is to show that, under a certain restriction on the elements f we can let $\varepsilon \to 0$ in (6.18).

This is the subject in the next section.

7 A two-parameter spectral theorem

Let $\{\Pi_s(\lambda), -\infty < \lambda < \infty\}$ be the spectral family of the self-adjoint operator $\overline{\Gamma}_s : H^{\otimes} \supset \mathcal{D}(\overline{\Gamma}_s) \to H^{\otimes}$, s = 1,2. Let

$$E(\lambda_1,\lambda_2) := \Pi_1(\lambda_1)\Pi_2(\lambda_2), \quad -\infty < \lambda_1, \lambda_2 < \infty$$

and finally let

$$\Pi(\lambda_1,\lambda_2) := E(\lambda_1,\lambda_2) \oplus E(\lambda_1,\lambda_2) : H \to H.$$

The most important results to date can now be stated:

Theorem 7.1: Let Condition 5.1 hold. Then

(a) $\Gamma : H \supset \mathcal{D}(\Gamma) \to H$ is essentially self-adjoint;

(b) $\overline{\Gamma}$ is abelian in the sense that its components $\overline{\Gamma}_1$ and $\overline{\Gamma}_2$ commute, i.e. their spectral families $\Pi_1(\lambda)$ and $\Pi_2(\mu)$ commute for all λ and μ.

(c) If $K \subset\subset (\mathbb{R} \setminus N_I) \times (\mathbb{R} \setminus N_J)$, ($N_I$ and N_J as in Condition 5.1) then, for all $f \in H$,

$$T\Pi(K)f = V \circ \overline{\Gamma}\Pi(K)f. \tag{7.1}$$

Proof: (a) is exactly Theorem 5.2;

(b) has been shown in §6.

(c) Firstly, we note that because K is compact it is sufficient to show that for every point $(\alpha,\beta) \in K$ there exists a rectangle $I : \lambda_1' < \lambda_1 \leq \lambda_1''$, $\lambda_2' < \lambda_2 \leq \lambda_2''$ with $(\alpha,\beta) \in \text{Int}(I)$ such that (7.1) holds with K = I. For then, by the Heine-Borel Theorem, K can be covered by a finite number of such rectangles $\{I_n\}_{n=1}^N$ such that every point of K is an interior point of at least one of the rectangles I_N. The K can be covered by a finite collection of sets $\{K_n\}_{n=1}^N$ such that $K = \bigcup_{n=1}^{N} K_n$ where $K_i \cap K_j = \varphi$, $i \neq j$ and $K_n \subset I_n$, n = 1,...,N. It follows that

$$\Pi(K) = \sum_{n=1}^{N} \Pi(K_n) \tag{7.2}$$

and

$$\Pi(I_n)\Pi(K_n) = \Pi(K_n) \quad \forall n = 1,\dots,N. \tag{7.3}$$

Thus, if (7.1) holds for $K = I_n$, i.e.

$$T\Pi(I_n)f = V \circ \overline{\Gamma}\Pi(I_n)f, \quad n = 1,\dots,N,$$

then it follows from (7.3) that

$$T\Pi(K_n)f = V \circ \overline{\Gamma}\Pi(K_n)f, \quad n = 1,\dots,N$$

so that (7.1) follows from (7.2).

Firstly, we show that, for any $(\alpha,\beta) \in K$, (5.4)(rs) implies that there exists a rectangle $I \ni (\alpha,\beta)$ such that, for $f_I := E(I)f$, $f \in H^{\otimes}$, there exists $c > 0$ such that

$$\|V_{\hat{r}\hat{s}}^{1/2} f_I\|^{\otimes} \leq c|||f_I|||^{\otimes}, \quad \hat{r} = 3-r, \quad \hat{s} = 3-s.$$

We consider (rs) = (11); the others follow in a similar manner.

By Lemma 2.8, (5.4)(11) implies that there exists an open interval $I' \ni \alpha$, and a constant $c > 0$, such that

$$\|E_1(I')u\|_1 \leq c|||E_1(I')u|||_{11} \quad \forall u \in H_1$$

where $\{E_1(\lambda), -\infty < \lambda < \infty\}$, is the spectral family for T_1. By considering the spectral family in $H^{\otimes}$ induced by $\{E_1(\lambda)\}$, we see that

$$\|E_1(I')u\|^{\otimes} \leq c|||E_1(\bar{I}')u|||_{11}^{\otimes} \quad \forall u \in H^{\otimes}$$

where

$$|||\cdot|||_{11}^{\otimes} := \sqrt{(\cdot,V_{11}\cdot)^{\otimes}} \ .$$

For $\mu \in I'$ let

$$K_\mu := \begin{cases} J \times (\alpha,\mu] & \text{for} \quad \mu > \alpha \\ J \times (\mu,\alpha] & \text{for} \quad \mu < \alpha \end{cases}$$

where J is an arbitrary compact set in $\mathbb{R}$.

Let $f_\mu := E(K_\mu)f$, $f \in H^\otimes$. Then, $f_\mu \in \mathcal{D}(\overline{\Gamma}_1) \cap \mathcal{D}(\overline{\Gamma}_2)$ and, from (6.18), we obtain

$$\begin{aligned} T_1 P_{22,\varepsilon} f_\mu &= P_{22,\varepsilon}(V_{11}\overline{\Gamma}_1 f_\mu + V_{12}\overline{\Gamma}_2 f_\mu) \\ &= V_{11}P_{22,\varepsilon}(\overline{\Gamma}_1 f_\mu - \overline{\Gamma}_2 f_\mu) + P_{22,\varepsilon}\overline{\Gamma}_2 f_\mu \\ &= V_{11}P_{22,\varepsilon}(\overline{\Gamma}_1 - \overline{\Gamma}_2) f_\mu + \int_{K_\mu} \lambda d(P_{22,\varepsilon} f_\lambda). \end{aligned}$$

Using integration by parts on the final term, we obtain

$$T_1 P_{22,\varepsilon} f_\mu = V_{11}P_{22,\varepsilon}(\overline{\Gamma}_1 - \overline{\Gamma}_2) f_\mu + \mu\,\mathrm{sgn}(\mu-\alpha)\ P_{22,\varepsilon} f_\mu - \int_{K_\mu} P_{22,\varepsilon} f_\lambda d\lambda.$$

Operating on this equation by $V_{22}^{1/2}$, and noting that $V_{22}^{1/2}\ T_1 \subset T_1\ V_{22}^{1/2}$, we obtain

$$\begin{aligned} &\|T_1 V_{22}^{1/2} P_{22,\varepsilon} f_\mu - \mu\,\mathrm{sgn}(\mu-\alpha) V_{22}^{1/2} P_{22,\varepsilon} f_\mu + \int_{K_\mu} V_{22}^{1/2} P_{22,\varepsilon} f_\lambda d\lambda\|^\otimes \\ &= \|V_{22}^{1/2} V_{11} P_{22,\varepsilon}(\overline{\Gamma}_1 - \overline{\Gamma}_2) f_\mu\|^\otimes \\ &\le |||P_{22,\varepsilon}(\overline{\Gamma}_1 - \overline{\Gamma}_2) f_\mu|||^\otimes \\ &\le |||(\overline{\Gamma}_1 - \overline{\Gamma}_2) f_\mu|||^\otimes \quad \text{(by Lemma 2.2(ii) since } P_{22,\varepsilon} \text{ commutes with } \Delta_o) \\ &\le |||\overline{\Gamma}_1 f_\mu|||^\otimes + |||\overline{\Gamma}_2 f_\mu|||^\otimes \\ &\le (\mathrm{Max}\ (|\alpha|,|\mu|) + \mathrm{Max}_{\lambda \in J}\ |\lambda|)\ |||f_\mu|||^\otimes \end{aligned}$$

i.e.

$$\|(T_1 - \mu\,\mathrm{sgn}(\mu-\alpha)) V_{22}^{1/2} P_{22,\varepsilon} f_\mu + \int_{K_\mu} V_{22}^{1/2} P_{22,\varepsilon} f_\lambda d\lambda\|^\otimes \le c'\,|||f_\mu|||^\otimes \qquad (7.4)$$

where $c' = \text{Max}\,(|\alpha|,|\mu|) + \text{Max}_{\lambda\in J}\,|\lambda|$.

Now fix μ_1 such that $[\alpha, \mu_1] \subset I'$ and let $\mu \in [\alpha, \mu_1]$. By Lemma 2.5 applied to the operator $T_1-\mu$ (which has spectral family $\{E_1(\lambda+\mu), \lambda \in \mathbb{R}\}$), there exist constants $a,b > 0$ such that

$$\begin{aligned} \|u\|^{\otimes} &\leq a\|(T_1-\mu)u\|^{\otimes} + b|||u|||_{11}^{\otimes} \\ &= a\|(T_1-\mu)u\|^{\otimes} + b\|V_{11}^{1/2}u\|^{\otimes} \end{aligned} \qquad \forall u \in \mathcal{D}(T_1). \tag{7.5}$$

Let $u = V_{22}^{1/2}P_{22,\varepsilon}f_\mu \in \mathcal{D}(T_1)$ in (7.5). Then, using (7.4), we get

$$\begin{aligned} \|V_{22}^{1/2}P_{22,\varepsilon}f_\mu\|^{\otimes} &\leq a\|(T_1-\mu)V_{22}^{1/2}P_{22,\varepsilon}f_\mu + \int_\alpha^\mu V_{22}^{1/2}P_{22,\varepsilon}f_\lambda d\lambda\|^{\otimes} + \\ &\quad + a\|\int_\alpha^\mu V_{22}^{1/2}P_{22,\varepsilon}f_\lambda d\lambda\|^{\otimes} + b\|V_{11}^{1/2}V_{22}^{1/2}P_{22,\varepsilon}f_\mu\|^{\otimes} \\ &\leq (ac'+b)\,|||f_\mu|||^{\otimes} + a(\mu-\alpha)m_\mu \end{aligned}$$

where $m_\mu := \sup_{\lambda\in[\alpha,\mu]} \|V_{22}^{1/2}P_{22,\varepsilon}f_\lambda\|^{\otimes}$. Since $\mu \in [\alpha,\mu_1] \subset I'$ and μ_1 is fixed, the constants a and b can be chosen independently of μ. Then, since $|||f_\mu|||^{\otimes}$, $\alpha \leq \mu$ is monotonically increasing with μ, it follows that

$$m_\mu \leq (ac'+b)\,|||f_\mu|||^{\otimes} + a(\mu-\alpha)m_\mu.$$

Choosing μ'' sufficiently close to α so that $a(\mu''-\alpha) \leq \frac{1}{2}$, we get

$$m_{\mu''} \leq (ac'+b)\,|||f_{\mu''}|||^{\otimes} + \frac{1}{2}\, m_{\mu''}$$

$$\Rightarrow \|P_{22,\varepsilon}V_{22}^{1/2}f_{\mu''}\|^{\otimes} \leq m_{\mu''} \leq 2(ac'+b)\,|||f_{\mu''}|||^{\otimes} \tag{7.6}$$

Now let $\varepsilon \to 0$. $P_{22,\varepsilon}V_{22}^{1/2}f_{\mu''} \xrightarrow{H^{\otimes}} V_{22}^{1/2}f_{\mu''}$. We require to show that $V_{22}^{1/2}f_{\mu''} \in H^{\otimes}$.

Choose a sequence $\{\varepsilon_n\}_{n=1}^{\infty}$, $\varepsilon_n \to 0$ as $n \to \infty$. From (7.6), $\{P_{22,\varepsilon_n}V_{22}^{1/2}f_{\mu''}\}_{n=1}^{\infty} \subset H^{\otimes}$ is a bounded sequence so that there exists a weakly convergent subsequence, say $\{g_n\}_{n=1}^{\infty}$: $g_n \xrightarrow{H^{\otimes}} h \in H^{\otimes}$, i.e.

$$(g_n,f)^{\otimes} \to (h,f)^{\otimes} \quad \forall f \in H^{\otimes}.$$

Then

$$[g_n-h,f]^{\otimes} = (g_n-h,\Delta_0 f)^{\otimes} \to 0 \quad \forall f \in \mathcal{H}^{\otimes}$$

i.e. g_n converges weakly to h in $\mathcal{H}^{\otimes}$.

But

$$g_n \xrightarrow{\mathcal{H}^{\otimes}} V_{22}^{1/2} f_{\mu''}$$

and "weak limit = strong limit" gives

$$V_{22}^{1/2} f_{\mu''} = h \in H^{\otimes}.$$

Further

$$g_n \xrightarrow{H^{\otimes}} h \Rightarrow \|h\|^{\otimes} \leq \sup \|g_n\|^{\otimes}$$

so that we deduce that

$$\|V_{22}^{1/2} f_{\mu''}\|^{\otimes} \leq 2(ac'+b)|||f_{\mu''}|||^{\otimes} \ . \tag{7.7}$$

Now, fix μ_0 so that $[\mu_0,\alpha] \subset I'$. Similarly, we can show that there exists $\mu' \in [\mu_0,\alpha)$ such that

$$\|V_{22}^{1/2} f_{\mu'}\|^{\otimes} \leq 2(ac'+b)|||f_{\mu'}|||^{\otimes} \tag{7.8}$$

Choosing $I = (\mu',\mu''] \times J$,

$$f_I = \Pi_1((\mu',\mu''])\Pi_2(J)f = f_{\mu''} + f_{\mu'}.$$

Then

$$\|V_{22}^{1/2}f_I\|^{\otimes^2} = \|V_{22}^{1/2}f_{\mu''} + V_{22}^{1/2}f_{\mu'}\|^{\otimes^2}$$

$$\leq 2\|V_{22}^{1/2}f_{\mu''}\|^{\otimes^2} + 2\|V_{22}^{1/2}f_{\mu'}\|^{\otimes^2}$$

$$\leq 8(ac'+b)^2(|||f_{\mu''}|||^{\otimes^2} + |||f_{\mu'}|||^{\otimes^2})$$

$$= 8(ac'+b)^2|||f_I|||^{\otimes^2}$$

which gives the result we wished to achieve.

Now, referring to (5.4), Condition 5.1 states that (11) or (21) holds and (12) or (22) holds, for each $(\alpha,\beta) \in K$. Suppose (11) and (12) hold. Then, as we have just shown, there exist rectangles I_1 and I_2 containing (α,β) such that

$$\|V_{2S}^{1/2}f_{I_S}\|^{\otimes} \leq c|||f_{I_S}|||^{\otimes}, \quad \forall f_{I_S} \in E(I_S)H^{\otimes}, \ s = 1,2,$$

for some constant $c > 0$. Then, for $I \subset I_1 \cap I_2$

$$\|V_{2S}^{1/2}f_I\|^{\otimes} \leq c|||f_I|||^{\otimes}, \quad \forall f_I \in E(I)H^{\otimes}, \ s = 1,2.$$

Now,

$$V_{21}^{1/2}V_{21}^{1/2} + V_{22}^{1/2}V_{22}^{1/2} = \mathrm{Id}$$

$$\Rightarrow \|f\|^{\otimes} \leq \|V_{21}^{1/2}f\|^{\otimes} + \|V_{22}^{1/2}f\|^{\otimes} \leq 2c|||f|||^{\otimes} \quad \forall f \in E(I)H^{\otimes}$$

so that

$$\|f\| \leq 2c|||f||| \quad \forall f \in \Pi(I)H.$$

For (21) and (22) holding, a similar estimate is proved in the same way.

Now suppose (11) and (22) hold. Then, as above, there exists a rectangle $I \ni (\alpha,\beta)$ and a constant $c > 0$ such that

$$\|V_{rr}^{1/2}f_I\|^{\otimes} \leq c|||f_I|||^{\otimes}, \quad r = 1,2 \tag{7.9}$$

Now it is easily seen that

$$\mathrm{Id} = \Delta_o + V_{12}V_{22} - V_{21}V_{11},$$

and

$$\|\Delta_o f\|^{\otimes} = |||\Delta_o^{1/2} f|||^{\otimes} \leq \sqrt{2}\,|||f|||^{\otimes} \quad \forall f \in H^{\otimes}.$$

Further,

$$\|V_{12}V_{22}f\|^{\otimes} = \|V_{12}V_{22}^{1/2}V_{22}^{1/2}f\|^{\otimes} \leq \|V_{22}^{1/2}f\|^{\otimes} \leq c\,|||f|||^{\otimes} \quad \forall f \in E(I)H^{\otimes}$$

and, similarly,

$$\|V_{21}V_{11}f\|^{\otimes} \leq c\,|||f|||^{\otimes} \quad \forall f \in E(I)H^{\otimes} .$$

Thus, we obtain

$$\|f\|^{\otimes} \leq (\sqrt{2} + 2c)\,|||f|||^{\otimes} \quad \forall f \in E(I)H^{\otimes}$$

so that

$$\|f\| \leq (\sqrt{2} + 2c)\,|||f||| \quad \forall f \in \Pi(I)H.$$

For (12) and (21) holding, a similar estimate is proved in the same way.

Thus Condition 5.1 implies that for any $(\alpha,\beta) \in K$ there exists a rectangle $I \ni (\alpha,\beta)$ such that

$$\|\Pi(I)f\| \leq c(I)\,|||\Pi(I)f||| \quad \forall f \in H. \tag{7.10}$$

In particular, note that $\Pi(I)f \in \mathrm{H} \quad \forall f \in H$. Now, consider (6.18):

$$TP_{\varepsilon}f = P_{\varepsilon}V \circ \overline{\Gamma}f \quad \forall f \in \mathcal{D}(\overline{\Gamma}).$$

For all $f \in H$, $\Pi(I)f \in \Pi(I)H \subset \mathcal{D}(\overline{\Gamma})$, so that

$$TP_\varepsilon \Pi(I)f = P_\varepsilon V \circ \bar{\Gamma}\Pi(I)f \quad \forall f \in H.$$

Now

$$P_\varepsilon \Pi(I)f \overset{H}{\to} \Pi(I)f \quad \text{as} \quad \varepsilon \to 0,$$

$$P_\varepsilon V \circ \bar{\Gamma}\Pi(I)f \overset{H}{\to} V \circ \bar{\Gamma}\Pi(I)f \quad \text{as} \quad \varepsilon \to 0,$$

and since T is a closed operator

$$\Pi(I)f \in \mathcal{D}(T)$$

and

$$T\Pi(I)f = V \circ \bar{\Gamma}\Pi(I)f \quad \forall f \in H.$$

From our initial discussion, the same result follows for any compact set $K \subset\subset (\mathbb{R}\setminus N_I) \times (\mathbb{R}\setminus N_J)$ i.e.

$$T\Pi(K)f = V \circ \bar{\Gamma}\Pi(K)f \quad \forall f \in H. \qquad \square \tag{7.11}$$

<u>Corollary 7.2:</u> Under the same assumptions as Theorem 7.1, for any $\Lambda \subset \mathbb{C} \times \mathbb{C}$, $K \subset\subset (\mathbb{R}\setminus N_I) \times (\mathbb{R}\setminus N_J)$,

$$(T - V \circ \Lambda)\Pi(K)f = V \circ (\bar{\Gamma} - \Lambda)\Pi(K)f \quad \forall f \in H. \tag{7.12}$$

<u>Proof:</u> Obvious. □

8 Localisation of the spectrum

To localize the spectrum of the two parameter problem it will be important to rotate the parameter space by a certain angle ϑ around a given point $\Lambda_o \in \mathbb{R}^2$. For this purpose we introduce the rotation matrix $\Theta(\vartheta)$:

$$\Theta(\vartheta) = \begin{pmatrix} \cos\vartheta & \sin\vartheta \\ -\sin\vartheta & \cos\vartheta \end{pmatrix}, \quad \vartheta \in (-\pi,\pi].$$

Now, consider

$$T_{\vartheta,\Lambda_o,ij} := (V \circ \Theta(\vartheta))^{-1}_{ij} \, (T - V \circ \Lambda_o)_i : H^{\otimes}_{\vartheta,ij} \supset D(T_i) \longrightarrow H^{\otimes}_{\vartheta,ij},$$

where

$$H^{\otimes}_{\vartheta,ij} := \overline{\{H^{\otimes},(\cdot,\cdot)^{\otimes}_{\vartheta,ij}\}}$$

$$(\cdot,\cdot)^{\otimes}_{\vartheta,ij} := (\cdot,|(V \circ \Theta(\vartheta))_{ij}|\cdot)^{\otimes}.$$

The norm will be denoted by $\|\cdot\|^{\otimes}_{\vartheta,ij}$. If $-\frac{\pi}{2} < \vartheta < 0$, then $H^{\otimes}_{\vartheta,11}$ and $H^{\otimes}_{\vartheta,22}$ are equivalent to $H^{\otimes}$, since $(V \circ \Theta(\vartheta))_{11} \gg 0$ and $(V \circ \Theta(\vartheta))_{22} \gg 0$ in this case.

If $0 < \vartheta < \frac{\pi}{2}$, then $H^{\otimes}_{\vartheta,12}$ and $H^{\otimes}_{\vartheta,21}$ are equivalent to $H^{\otimes}$, since in this case we have $(V \circ \Theta(\vartheta))_{12} \gg 0$ and $(V \circ \Theta(\vartheta))_{21} \gg 0$. Let $\{E_{\vartheta,\Lambda_o,ij}(\lambda),\ \lambda \in \mathbb{R}\}$ be the spectral family of $T_{\vartheta,\Lambda_o,ij}$. We first choose $\vartheta = -\frac{\pi}{4}$. If $\{E_{\Lambda_o,i}(\lambda),\ \lambda \in \mathbb{R}\}$ is the spectral family of $(T - V \circ \Lambda_o)_i$ we have

$$E_{\Lambda_o,i}(\lambda) = 2^{-1/4} E_{-\frac{\pi}{4},\Lambda_o,ii}(2^{-1/2}\lambda), \qquad i = 1,2.$$

Note that multiplication by $2^{-1/4}$ is a unitary operation between $H^{\otimes}_{-\frac{\pi}{4},ii}$ and $H^{\otimes}$. We define

$$E_{o,\Lambda_o}(\lambda,\mu) := E_{\Lambda_o,1}(\lambda)\, E_{\Lambda_o,2}(\mu), \quad (\lambda,\mu) \in \mathbb{R}^2,$$
$$E_{\Lambda_o}(\Lambda) := E_{o,\Lambda_o}(\Lambda) \oplus E_{o,\Lambda_o}(\Lambda), \quad \Lambda \in \mathbb{R}^2. \tag{8.1}$$

<u>Theorem 8.1:</u> Let K be a compact subset of $(\mathbb{R}\setminus N_I) \times (\mathbb{R}\setminus N_J)$ with

$$\Lambda_o \in K \subset\subset \Lambda_o + \Theta(-\tfrac{\pi}{4})I,$$

where $I \subset \mathbb{R}^2$ is a rectangle containing the origin. Then

$$\|(\Pi(K) - E_{\Lambda_o}(I)\Pi(K))f\| \le c(I,K)\ |||\Pi(K)f||| \tag{8.2}$$

where

$$c(I,K) = 4\sqrt{2}\ c(K)\ \frac{\max\limits_{\Lambda\in K}|\Lambda_o-\Lambda|}{\min\limits_{(\alpha,\beta)\in\partial I} \min(|\alpha|,|\beta|)},$$

and c(K) is the constant appearing in the proof of Theorem 7.1.

<u>Proof:</u> We have from (7.12) that

$$(T-V \circ \Lambda_o)\ \Pi(K) = V \circ (\overline{\Gamma} - \Lambda_o)\ \Pi(K). \tag{8.3}$$

Multiplying this by $(Id-E_{\Lambda_o}(I))$ we obtain

$$(Id-E_{\Lambda_o}(I))\ (T-V \circ \Lambda_o)\ \Pi(K) = (Id-E_{\Lambda_o}(I))\ V \circ (\overline{\Gamma} - \Lambda_o)\ \Pi(K)$$

$$\Leftrightarrow (T-V \circ \Lambda_o)(Id-E_{\Lambda_o}(I))\ \Pi(K) = (Id-E_{\Lambda_o}(I))\ V \circ (\overline{\Gamma} - \Lambda_o)\ \Pi(K)$$

$$\Leftrightarrow (Id-E_{\Lambda_o}(I))\Pi(K) = (T-V \circ \Lambda_o)^{-1}(Id-E_{\Lambda_o}(I))V \circ (\overline{\Gamma} - \Lambda_o)\Pi(K).$$

This leads to

$$\|(Id-E_{\Lambda_o}(I))\Pi(K)f\| \le \frac{\sqrt{2}}{\min\limits_{(\alpha,\beta)\in\partial I} \min(|\alpha|,|\beta|)} \|V \circ (\bar{\Gamma} - \Lambda_o)\Pi(K)f\|$$

$$\le \frac{2\sqrt{2}\ c(K)}{\min\limits_{(\alpha,\beta)\in\partial I} \min(|\alpha|,|\beta|)} |||(\bar{\Gamma} - \Lambda_o)\Pi(K)f|||$$

$$\le \frac{4\sqrt{2}\ c(K)}{\min\limits_{(\alpha,\beta)\in\partial I} \min(|\alpha|,|\beta|)} \max_{\Lambda\in K} |\Lambda-\Lambda_o|\ |||\Pi(K)f|||. \qquad \square$$

As a consequence of Theorem 8.1 we obtain

Theorem 8.2: If $(T-V \circ \Lambda_o)$, $\Lambda_o \in \mathbb{R}^2$, has at least one boundedly invertible component then

$$\Pi(K) = 0$$

for some open rectangle $K \subset \mathbb{R}^2$ such that $\Lambda_o \in K$.

Proof: Without loss of generality we may assume that the first component $(T-V \circ \Lambda_o)_1$ of $T-V \circ \Lambda_o$ is boundedly invertible.

Firstly, we show that $\Lambda_o = (\lambda_{o,1}, \lambda_{o,2})$ is a regular point, i.e. $\Lambda_o \in (\mathbb{R}\setminus N_I) \times (\mathbb{R}\setminus N_J)$. Observe that by the assumptions on V we have

$$T_1 - V_{11}\lambda_{o,1} - V_{12}\lambda_{o,2} = T_1 - V_{11}(\lambda_{o,1} - \lambda_{o,2}) - \lambda_{o,2}.$$

With $\alpha = \lambda_{o,1} - \lambda_{o,2}$, let $\{E^\alpha(\lambda) := E^\alpha_{11}(\lambda),\ \lambda \in \mathbb{R}\}$ be the spectral family of $T_1 - V_{11}(\lambda_{o,1} - \lambda_{o,2})$. Then $E^\alpha(\lambda) = 0$ near $\lambda_{o,2}$. In particular, there exists some interval $I \subset \mathbb{R}$ containing $\lambda_{o,2}$ such that

$$\|E^\alpha(I)u\|^\otimes \le c|||E^\alpha(I)u|||^\otimes_{11} \qquad \forall\ u \in H^\otimes. \tag{8.4}$$

According to Lemma 2.10 we have

$$(T_1 - V_{11}(\lambda_{o,1}-\lambda_{o,2}) \pm i\ V_{11} - V_{12}\ \lambda_{o,2})^{-1} \in B(H^\otimes)$$

i.e. $(T_1 - (\lambda_{o,1}- \lambda_{o,2}) \pm i\ V_{11} - V_{12}(\lambda_{o,2} - (\lambda_{o,1}- \lambda_{o,2})))^{-1} \in B(H^\otimes)$. (8.5)

Then, according to Lemma 2.8 [with $\alpha = 0$ and $E_1(\lambda + (\lambda_{o,1} - \lambda_{o,2}))$ as spectral family of $(T_1 - (\lambda_{o,1} - \lambda_{o,2})]$ we have an open interval I and a constant $c > 0$ such that

$$\|E_1(I + (\lambda_{o,1} - \lambda_{o,2}))u\|^{\otimes} \leq c\,|||E_1(I + (\lambda_{o,1} - \lambda_{o,2}))u|||^{\otimes}_{11} \quad \forall u \in H^{\otimes}$$

where

$$\lambda_{o,2} - (\lambda_{o,1} - \lambda_{o,2}) \in I,$$

i.e. $\lambda_{o,2} \in I + (\lambda_{o,1} - \lambda_{o,2}) =: I'$.

Thus

$$\|E_1(I')u\|^{\otimes} \leq c\,|||E_1(I')u|||^{\otimes}_{11} \qquad \forall\, u \in H^{\otimes}.$$

Again applying Lemma 2.10, we have the existence of

$$(T_1 \pm V_{11}i - V_{12}\,\lambda_{o,2})^{-1} \in B(H^{\otimes}).$$

Similarly, we obtain the existence of

$$(T_1 - V_{11}\,\lambda_{o,1} \pm i\,V_{12})^{-1} \in B(H^{\otimes}).$$

These are exactly the requirements appearing in (5.1) and (5.2).

Now, it follows from our assumptions that there exist $\alpha,\beta > 0$ such that

$$E_{\Lambda_o}((-\alpha,\alpha] \times (-\beta,\beta]) = 0.$$

Then, from (8.2) with $I = (-\alpha,\alpha] \times (-\beta,\beta]$, we have

$$\|\Pi(K)f\| \leq c(I,K)\,|||\Pi(K)f||| \leq c'(I,K)\|\Pi(K)f\| \tag{8.6}$$

for any regular domain $K \subset\subset (\mathbb{R}\setminus N_I) \times (\mathbb{R}\setminus N_J)$ with $\Lambda_o \in K \subset\subset \Lambda_o + \Theta(-\frac{\pi}{4})I$. Since $c'(I,K) \longrightarrow 0$ if diameter of K tends to 0, it follows from (8.6) that $\Pi(K)f \equiv 0$ for some open rectangle K. □

The natural definition of the spectrum of the two-parameter problem is the set of points

$$\{\Lambda \in \mathbb{C}^2 \mid \text{neither component of } (T-V \circ \Lambda) \text{ has a bounded inverse}\}.$$

Theorem 8.2 shows that this set is related to the support of the spectral measure $\Pi(\cdot)$ i.e. the set

$$\{\Lambda \in \mathbb{R}^2 \mid \Pi(K) \neq 0 \quad \forall \text{ open intervals } K \text{ such that } \Lambda \in K\}.$$

9 Some further lemmas

In this section we collect some further results which are required to complete the spectral representation theorem. The results in this section are valid in an arbitrary separable Hilbert space. Therefore, as in Section two, we will use the notation H, $\mathcal{H}$, etc., in a general sense rather than to denote the specific spaces under consideration in Sections three to eight.

The first two results are concerned with a Neumann series argument and its refinements for solving certain operator equations.

Lemma 9.1: Let H be a Hilbert space with inner product $(\cdot,\cdot)$, and norm $\|\cdot\|$. Let $V : H \supset \mathcal{D}(V) \to H$ satisfy $V >> 0$, and let $\mathcal{H} := \overline{\{\mathcal{D}(V), [\cdot,\cdot]\}}$ where

$$[\cdot,\cdot] := (\cdot,V\cdot) \quad \text{and} \quad |||\cdot||| := \sqrt{[\cdot,\cdot]}.$$

Let $P \in B(\mathcal{H})$ and $Q \in B(H)$ with $|||P||| = \gamma_P$ and $\|Q\| = \gamma_Q$ satisfying

$$\gamma_P \, \gamma_Q < 1. \tag{9.1}$$

Then, for every bounded linear mapping $A : H \to \mathcal{H}$ (i.e. $\exists\, c_A > 0$ such that $|||Av||| \leq c_A\|v\| \quad \forall v \in H$), the operator equation

$$X - PXQ = A \tag{9.2}$$

is uniquely solvable. The solution $X : H \to \mathcal{H}$ satisfies

$$|||Xv||| \leq \frac{c_A}{1-\gamma_P\gamma_Q}\|v\| \qquad \forall v \in H \tag{9.3}$$

and can be represented as

$$X = \sum_{n=0}^{\infty} P^n AQ^n \tag{9.4}$$

in the sense of strong convergence.

Proof: If X satisfies (9.2), then by iteration we obtain

$$X = \sum_{\nu=0}^{n-1} P^{\nu} AQ^{\nu} + P^{n} X Q^{n}. \tag{9.5}$$

Now, since

$$|||P^{\nu}A\, Q^{\nu}v||| \le \gamma_P^{\nu}\, \gamma_Q^{\nu}\, c_A\|v\|, \tag{9.6}$$

we see that $\sum_{\nu=0}^{\infty} P^{\nu}A\, Q^{\nu}$ exists strongly and is a solution of (9.2). In particular, we have $P^n\, X\, Q^n \to 0$ as $n \to \infty$. This yields the existence and the representation (9.4) of the solution of (9.2). Then

$$|||Xv||| = |||\sum_{\nu=0}^{\infty} P^{\nu}\, A\, Q^{\nu}v||| \le \sum_{\nu=0}^{\infty} \gamma_P^{\nu}\, \gamma_Q^{\nu}\, c_A\|v\| = \frac{c_A\|v\|}{1-\gamma_P\gamma_Q} \quad \forall v \in H \qquad \square$$

We need a refinement of Lemma 9.1 for the case $\gamma_P\, \gamma_Q = 1$.

Lemma 9.2: Let $\gamma_P\, \gamma_Q \le 1$ and in addition to the assumption of the previous lemma suppose that P and Q are self-adjoint and 1 is not an eigenvalue of $(\gamma_Q Q)^2$ so that $W := (\mathrm{Id} - (\gamma_Q Q)^2)^{-1}$ exists as a self-adjoint operator. Then there exists a solution $X : \mathcal{D}(X) \subset H \to H$ of equation (9.2) with

$$\mathcal{D}(X) = \bigcup_{\varkappa\in(0,\frac{1}{2}]} \mathcal{D}(W^{1+\varkappa}).$$

Moreover, representation (9.4) holds on $\mathcal{D}(X)$ and we have the estimate

$$|||Xv||| \le c_A\, c(\varkappa)\|W^{1+\varkappa}v\| \quad \forall v \in \mathcal{D}(W^{1+\varkappa}), \quad \varkappa \in (0,\tfrac{1}{2}], \tag{9.7}$$

where $c(\varkappa)$ may be chosen as

$$c(\varkappa) = 2(1 + 8\,(1-\varkappa)\,\Gamma(\tfrac{1}{2} + \varkappa)\sqrt{\zeta(1 + 2\varkappa)}) \tag{9.8}$$

and Γ denotes the Gamma function and ζ denotes the Riemann zeta-function. If Y is a solution of (9.2) with $\mathcal{D}(Y) \supset \mathcal{D}(W^{\varkappa_o})$ and

$$|||Yv||| \le c_Y\|W^{\varkappa_o}\, v\| \quad \forall v \in \mathcal{D}(W^{\varkappa_o}), \tag{9.9}$$

for some $\varkappa_0 \in \mathbb{R}$ then

$$X = Y \text{ on } \mathcal{D}(W^{\varkappa_0}) \cap \mathcal{D}(X). \tag{9.10}$$

Proof: Because of Lemma 9.1 we may restrict our attention to the case $\gamma_P \gamma_Q = 1$. Without loss of generality we may further assume $\gamma_P = \gamma_Q = 1$ (replace P by γ_P^{-1} P and Q by $\gamma_Q Q$).

Firstly, we show the existence of the limit

$$Xv = \sum_{n=0}^{\infty} P^n A Q^n v \quad \forall v \in \mathcal{D}(W^{1+\varkappa}), \quad \varkappa \in (0,\tfrac{1}{2}]. \tag{9.11}$$

This shall be achieved by considering

$$X_\pm v = \sum_{n=0}^{\infty} (P_\pm)^n A Q^n v \quad \forall v \in \mathcal{D}(W^{1+\varkappa}), \quad \varkappa \in (0,\tfrac{1}{2}], \tag{9.12}$$

separately, where $P_\pm$ denotes the positive and negative part of P respectively, so that $P = P_+ + P_-$. As a preparatory observation we note that for the factorial function and the binomial coefficient we have the following relations

$$(-1)^\nu \binom{-(1/2+\varkappa)}{\nu} = \frac{(\frac{1}{2}+\varkappa)_\nu}{\nu!} > 0 \text{ for } \varkappa > 0 \text{ and } \nu \in \mathbb{N},$$

$$(\tfrac{1}{2}+\varkappa)_\nu \mu! \le (\tfrac{1}{2}+\varkappa)_\mu \nu! \quad \text{for } \varkappa \in (0,\tfrac{1}{2}] \text{ and } \mu,\nu \in \mathbb{N} \text{ with } \mu < \nu.$$

Thus, for $x \in \mathbb{R}$, $|x| < 1$, and $\varkappa \in (0,\frac{1}{2}]$ we have the estimates

$$|\sum_{\nu=n}^{\infty} \binom{-(1/2+\varkappa)}{\nu} x^\nu| \le |x|^n \sum_{\nu=0}^{\infty} (-1)^\nu \binom{-(1/2+\varkappa)}{\nu} |x|^\nu$$
$$= |x|^n (1-|x|)^{-(1/2+\varkappa)} \le 2|x|^n (1-x^2)^{-(1/2+\varkappa)},$$

and

$$|\sum_{\nu=n}^{\infty} (-1)^\nu \binom{-(1/2+\varkappa)}{\nu} x^\nu| \le 2\,|x|^n (1-x^2)^{-(1/2+\varkappa)}.$$

By applying the spectral theorem we obtain from these estimates that

$$Z_n^- := \sum_{\nu=n}^{\infty} \binom{-(1/2+\varkappa)}{\nu} Q^\nu, \qquad Z_n^+ := \sum_{\nu=n}^{\infty} (-1)^\nu \binom{-(1/2+\varkappa)}{\nu} Q^\nu, \tag{9.13}$$

are well defined, essentially self-adjoint operators on $\mathcal{D}(W^{1/2+\varkappa})$ and we have

$$\|Z_n^\pm v\| \leq 2\| |Q|^n W^{1/2+\varkappa} v\| \qquad \forall v \in \mathcal{D}(W^{1/2+\varkappa}). \tag{9.14}$$

It is sufficient to consider X_+ since X_- can be treated analogously. For $v \in \mathcal{D}(W^{1/2+\varkappa})$ we have

$$Q^n = a_n^+ (Z_{n+1}^+ - Z_n^+)$$

where

$$a_n^+ = (-1)^{n+1} \left\{ \binom{-(1/2+\varkappa)}{n} \right\}^{-1}.$$

So, for $0 \leq m_o < n_o$ we may write

$$\left.\begin{aligned} \sum_{n=m_o}^{n_o} (P_+)^n A Q^n v &= -a_{m_o}^+ (P_+)^{m_o} A Z_{m_o}^+ v + a_{n_o}^+ (P_+)^{n_o} A Z_{n_o+1}^+ v \\ &\quad + \sum_{n=m_o+1}^{n_o} [a_{n-1}^+ (P_+)^{n-1} - a_n^+ (P_+)^n] A Z_n^+ v \\ &= -a_{m_o}^+ (P_+)^{m_o} A Z_{m_o}^+ v + a_{n_o}^+ (P_+)^{n_o} A Z_{n_o+1}^+ v \\ &\quad + \sum_{n=m_o}^{n_o-1} a_n^+ (P_+)^n (1-P_+) A Z_{n+1}^+ v \\ &\quad + \sum_{n=m_o}^{n_o-1} a_n^+ \left(1 - \frac{a_{n+1}^+}{a_n^+}\right) (P_+)^{n+1} A Z_{n+1}^+ v . \end{aligned}\right\} \tag{9.15}$$

By making use of

$$|a_n^+| \leq c_1(\varkappa)\ (n+1)^{1/2-\varkappa}, \tag{9.16}$$

where $c_1(\varkappa)$ may be expressed in terms of a Γ-function as

$$c_1(\varkappa) = 2\Gamma(\tfrac{1}{2} + \varkappa),$$

we can estimate the first two terms on the right-hand side of (9.15) as follows. If $m_0 = 0$ then we use

$$|||a_0^+ \ A\ Z_0^+\ v||| \leq 2\ c_A\ \|W^{1/2+\varkappa}\ v\| \leq 2\ c_A\ \|W^{1+\varkappa}\ v\|.$$

The second term can be estimated using (9.16) and (9.14):

$$\begin{aligned} |||a_{n_0}^+\ (P_+)^{n_0}\ A\ Z_{n_0+1}^+\ v||| &\leq 2\ c_1(\varkappa)\ (n_0+1)^{1/2-\varkappa}\ c_A\||Q|^{n_0+1}\ W^{1/2+\varkappa}\ v\| \\ &\leq 2\ c_1(\varkappa)\ (n_0+1)^{1/2-\varkappa}(n_0+2)^{-1/2}\ c_A\|W^{1+\varkappa}\ v\|; \end{aligned} \tag{9.17}$$

the latter estimate is based on

$$|x|^n\ (1-x^2)^{1/2} \leq (n+1)^{-1/2},\ x \in \mathbb{R},\ |x| \leq 1,\ n \in \mathbb{N}. \tag{9.18}$$

The same result holds for $m_0 > 0$ in place of n_0. Estimate (9.17) shows in particular that, provided $\varkappa > 0$,

$$\lim_{n\to\infty} |||a_n^+\ (P_+)^n\ A\ Z_{n+1}^+\ v||| = 0.$$

Now, let us consider the first sum term in (9.15). Again using (9.14) and (9.16) we have for $u \in H$, $v \in H$

$$|[u, \sum_{n=m_o}^{n_o-1} a_n^+ (P_+)^n (1-P_+)A\, Z_{n+1}^+ v]|^2$$

$$\leq 4\, c_A^2 \sum_{n=m_o}^{n_o-1} |a_n^+|^2 |||(P_+)^n (1-P_+)\, u|||^2 \sum_{n=m_o}^{n_o-1} \||Q|^{n+1} W^{1/2+\varkappa} v\|^2$$

$$\leq 4\, c_A^2 (c_1(\varkappa))^2 \sum_{n=m_o}^{n_o-1} (n+1)^{-1-2\varkappa} |||u|||^2 (W^{1/2+\varkappa} v, W\, W^{1/2+\varkappa} v) \tag{9.19}$$

$$\leq 4\, c_A^2 (c_1(\varkappa))^2 \sum_{n=m_o}^{n_o-1} (n+1)^{-1-2\varkappa} |||u|||^2 \|W^{1+\varkappa} v\|^2 .$$

Here (9.18), (9.14) have been used. Estimate (9.19) implies in particular that

$$|||\sum_{n=m_o}^{n_o-1} a_n^+ (P_+)^n (1-P_+)\, A\, Z_{n+1}^+ v|||$$

$$\leq 2\, c_A\, c_1(\varkappa) \sqrt{\sum_{n=m_o}^{n_o-1} (n+1)^{-1-2\varkappa}}\, \|W^{1+\varkappa} v\| \tag{9.20}$$

$\rightarrow 0$ as $m_o \rightarrow \infty$ provided $\varkappa > 0$.

Similarly one can deal with the remaining sum term in (9.15) taking into account that

$$(1 - \frac{a_{n+1}^+}{a_n^+}) = \frac{\varkappa - 1/2}{\frac{1}{2} + \varkappa + n} .$$

For $\varkappa \leq 1/2$, we obtain

$$|||\sum_{n=m_o}^{n_o-1} a_n^+ \left(1 - \frac{a_{n+1}^+}{a_n^+}\right) (P_+)^{n+1} A\, Z_{n+1}^+ v|||$$

$$\leq 2(1-2\varkappa)\, c_A\, c_1(\varkappa) \sqrt{\sum_{n=m_o}^{n_o-1} (n+1)^{-1-2\varkappa}}\, \|W^{1+\varkappa} v\|$$

$\rightarrow 0$ as $m_o \rightarrow \infty$ provided $\varkappa > 0$.

By replacing a_n^+, Z_n^+ with a_n^-, Z_n^- we obtain the corresponding result for X_-

instead of X_+. Thus we have the desired convergence in H of the series in (9.11). Letting $m_0 = 0$ and $n_0 \to \infty$ in (9.15) and its equivalent for P_- we get an estimate for the limit X:

$$|||Xv||| \leq 2\ c_A(1 + 4(1-\varkappa)\ c_1(\varkappa) \sqrt{\zeta(1 + 2\varkappa)})\ \|W^{1+\varkappa}\ v\| \tag{9.21}$$

for all $v \in \mathcal{D}(W^{1+\varkappa})$. In order to prove the first part of the lemma it only remains to show that X is a solution. But this is easily established:

$$P\ X\ Q\ v = \sum_{n=1}^{\infty} P^n\ A\ Q^n\ v = (X-A)v$$

for any $v \in \mathcal{D}(W^{1+\varkappa})$. Moreover, according to (9.21) the desired estimate holds. Now, let Y be a solution of (9.2) with $\mathcal{D}(Y) \supset \mathcal{D}(W^{\varkappa_0})$ and

$$|||Yv||| \leq c_Y\ \|W^{\varkappa_0}\ v\|\ \forall v \in \mathcal{D}(W^{\varkappa_0}), \tag{9.22}$$

for some $\varkappa_0 \in \mathbb{R}$. Then we have by iterating equation (9.2) the relation

$$Yv = \sum_{\nu=0}^{n-1} P^\nu\ A\ Q^\nu\ v + P^n\ Y\ Q^n\ v \qquad \forall\ v \in \mathcal{D}(W^{\varkappa_0}).$$

For the remainder term the estimate

$$|||P^n\ Y\ Q^n\ v||| \leq |||Y\ Q^n\ v||| \leq c_Y\ \|W^{\varkappa_0}\ Q^n\ v\| = c_Y\ \||Q|^n\ W^{\varkappa_0}\ v\| \tag{9.23}$$

holds for all $v \in \mathcal{D}(W^{\varkappa_0})$. Since by assumption $\lambda = 1$ is not an eigenvalue of $|Q|$ we get from (9.23) that the remainder term tends to zero as $n \to \infty$. □

Lemma 9.3: Let H be a Hilbert space with inner product $(\cdot,\cdot)$ and norm $\|\cdot\|$. Let $V \in B(H)$ satisfy $V > 0$ and let $H := \{H, [\cdot,\cdot]\}$ where

$$[\cdot,\cdot] := (\cdot,V\cdot) \quad \text{and} \quad |||\cdot||| := \sqrt{[\cdot,\cdot]}.$$

Let P be an orthogonal projection on H such that $R(P)$ is closed in H. Then

$$V_P := PVP \quad : \quad R(P) \to R(P)$$

is a bijection so that V_P^{-1} exists and

$$P_V := (V_P)^{-1}PV \quad : \quad H \to H$$

is an H-orthogonal projection on $R(P)$.

We prove the lemma in the following stages:

(1) Firstly we show that $PVP : R(P) \to R(P)$ is strongly positive definite. For all $v \in R(P)$,

$$\begin{aligned}(PVPv, v) &= (VPv, Pv) \\ &= (Vv, v) \qquad \text{since} \quad Pv = v \\ &> 0 \qquad\quad \text{since} \quad V > 0.\end{aligned}$$

Now suppose there exists a sequence of unit vectors $\{u_n\}_{n=1}^{\infty} \subset R(P)$. $\|u_n\| = 1$, such that $(Vu_n, u_n) \to 0$ as $n \to \infty$. Then

$$|||u_n|||^2 = (Vu_n, u_n) \to 0 \quad \text{as} \quad n \to \infty,$$

so that $u_n \xrightarrow{H} 0$. Choose a subsequence, which we shall relabel $\{u_n\}_{n=1}^{\infty}$ such that

$$|||u_n||| \leq \frac{1}{2^n} \qquad \forall n,$$

and consider

$$v_N := \sum_{n=1}^{N} u_n \in R(P) \quad \forall\, N \in \mathbb{N}.$$

Then,

$$|||v_{N+P} - v_N||| \leq \sum_{n=N+1}^{N+p} |||u_n||| \leq \sum_{n=N+1}^{N+p} \frac{1}{2^n} < \frac{1}{2^N}$$

so that $\{v_N\}_{N=1}^{\infty} \subset R(P)$ is Cauchy in H. Since $R(P)$ is assumed to be closed in H, it follows that

$$v := \lim_{N\to\infty} v_N = \sum_{n=1}^{\infty} u_n \in R(P);$$

but this is a contradiction, since $\sum_{n=1}^{\infty} u_n$ is not convergent in H and $R(P) \subset H$.

Thus, we can conclude that $V_P := PVP \gg 0$ on $R(P)$, i.e. $\exists c > 0$ such that $(V_P v, v) > c\|v\| \quad \forall v \in R(P)$.

(2) It follows immediately that $V_P : R(P) \to R(P)$ is self-adjoint on $\{R(P), (\cdot,\cdot)\}$ and is a bijection so that $V_P^{-1} : R(P) \to R(P)$ is bounded, and

$$P_V := (V_P)^{-1}PV : H \to H$$

is everywhere defined (note that, by Lemma 2.2(iii) $VH \subset H$).

(3) $P_V v = v$ for all $v \in R(P)$, and $P_V^2 = P_V$ on H: Let $v \in R(P)$. Then, for all $u \in R(P)$

$$\begin{aligned}[u,P_V v] &= (u,V_P P_V v)\\ &= (u,P\ V\ v)\\ &= (u,\ V\ v)\\ &= [u,v].\end{aligned}$$

Consequently $P_V\ v = v$ for all $v \in R(P)$, and, since $P_V\ u \in R(P)$ for all $u \in H$, we obtain $P_V\ P_V = P_V$.

(4) $R(P)^{\perp H} = N(P_V)$:

Let $v \in H$ and for all $u \in R(P)$ suppose $[u,v] = 0$. Then

$$(u,\ Vv) = 0 \qquad \forall u \in R(P),$$

$$\Leftrightarrow (u,PV\ v) = 0 \ \forall u \in H. \tag{9.24}$$

Note that $Vv \in H$ by Lemma 2.2(iii). Thus by (9.24) we have

$$PV\ v = 0.$$

Consequently

$$P_V\ v := V_P^{-1}\ P\ V\ v = 0$$

so that

$$\mathcal{R}(P)^{\perp H} \subset \mathcal{N}(P_V).$$

The reverse inclusion is obtained in a similar fashion.

(5) P_V is self-adjoint on $\mathcal{H}$:

$$\begin{aligned}[u, P_V\, v] &= (Vu, P_V\, v) \\ &= (P\ V\ u,\ V_P^{-1}\ P\ V\ v) \\ &= (V_P^{-1}\ P\ V\ u, P\ V\ v) \\ &= (P_V\ u,\ V\ v) \\ &= [P_V\ u, v] \quad \text{for all } u, v \in \mathcal{H}.\end{aligned}$$

From what has been shown it is clear that $P_V : \mathcal{H} \to \mathcal{H}$ is an orthogonal projection on $\mathcal{R}(P_V)$. Moreover,

$$\mathcal{R}(P_V) = \mathcal{N}(P_V)^{\perp H} = \mathcal{N}(P)^{\perp} = \mathcal{R}(P). \qquad \square$$

Now, we turn our attention to properties of spectral measures to be employed later.

<u>Lemma 9.4:</u> Let $\{\varphi_i\}_{i=1}^{\infty}$ be a complete orthonormal system in a Hilbert space H, and let $\{E(\lambda),\ \lambda \in \mathbb{R}\}$ be a spectral family. Then, for $\varphi \in H$, $\|E(\lambda)\varphi\|^2$ is absolutely continuous with respect to

$$\rho(\lambda) = \sum_{k=1}^{\infty} k^{-2}\ (\varphi_k, E(\lambda)\varphi_k). \tag{9.25}$$

<u>Proof:</u> Let $\varphi = \sum_{i=1}^{N} a_i\ \varphi_i \in H$, where N is finite. Then

$$\|E(I)\varphi\|^2 = \sum_{i=1}^{N} |a_i|^2 \|E(I)\varphi_i\|^2 \le (N \max |a_i|)^2 \cdot \rho(I).$$

This shows the absolute continuity of $\|E(\lambda)\varphi\|^2$ w.r.t. $\rho(\lambda)$. An approximation argument gives the result for general $\varphi \in H$. $\square$

Lemma 9.5: The set

$$H_\rho := \{v \in H \mid \sup_{\lambda\in\mathbb{R}} \frac{d\|E(\lambda)\, v\|^2}{d\rho(\lambda)} < \infty\} \tag{9.26}$$

is dense in H.

Proof: Let $\{\varphi_i\}_{i=1}^{\infty}$ be a complete orthonormal system in H.

If $u = \sum_{i=1}^{N} a_i\, \varphi_i$, $a_i \in \mathbb{C}$, $i = 1,\dots,N$, then u is in H_ρ. Since the orthonormal system is complete in H the result follows. □

Lemma 9.6: Let $\rho(\lambda)$, $\lambda \in [\lambda_1,\lambda_2]$ be monotonically non-decreasing and right-continuous. Then for any $[\lambda_1',\lambda_2'] \subset\subset (\lambda_1,\lambda_2)$ there exists $\lambda_0 \in [\lambda_1',\lambda_2']$ such that

$$\gamma(\lambda_0) := \sup_{\mu\in[\lambda_1,\lambda_2]} \frac{\rho(\lambda_0) - \rho(\mu)}{\lambda_0 - \mu} \le 2\, \frac{\rho(\lambda_2) - \rho(\lambda_1)}{\lambda_2' - \lambda_1'} . \tag{9.27}$$

Proof: The proof is by contradiction. Let

$$\gamma_-(\lambda) := \sup_{\mu\in[\lambda_1,\lambda)} \frac{\rho(\lambda) - \rho(\mu)}{\lambda - \mu} ,$$

$$\gamma_+(\lambda) := \sup_{\mu\in(\lambda,\lambda_2]} \frac{\rho(\lambda) - \rho(\mu)}{\lambda - \mu} , \tag{9.28}$$

and define

$$m_\pm := \{\lambda \in [\lambda_1',\lambda_2') \mid \gamma_\pm(\lambda) > \zeta\}, \text{ where } \zeta := 2\, \frac{\rho(\lambda_2) - \rho(\lambda_1)}{\lambda_2' - \lambda_1'} .$$

Since ρ is right-continuous, we have that $m_\pm$ is the union of at most countably many disjoint, right-open intervals.

Assume $[\lambda_1',\lambda_2') = m_+ \cup m_-$ so that $\gamma(\lambda) > \zeta$ for all $\lambda \in [\lambda_1',\lambda_2')$.

Note that $\gamma(\lambda) > \zeta$ implies that $\gamma_+(\lambda) > \zeta$ or $\gamma_-(\lambda) > \zeta$. We first observe that $m_+ \cap (\lambda_1,\lambda_2)$ is open. This can be seen as follows.

Let $\lambda \in m_+ \cap (\lambda_1,\lambda_2)$. Then there exists $\mu > \lambda$ such that

$$\frac{\rho(\lambda) - \rho(\mu)}{\lambda - \mu} > \zeta,$$

and so

$$\frac{\rho(\lambda+) - \rho(\mu)}{\lambda - \mu} > \zeta,$$

by the right-continuity. Since, on the other hand

$$\frac{\rho(\lambda-) - \rho(\mu)}{\lambda - \mu} \geq \frac{\rho(\lambda) - \rho(\mu)}{\lambda - \mu},$$

by the monotonicity of ρ, we see that in fact $m_+ \cap (\lambda_1,\lambda_2)$ is open.

To proceed with the contradiction argument we distinguish two cases:

1. $m_+ \cap m_- = \emptyset$,
2. $m_+ \cap m_- \neq \emptyset$.

Case 1: Since $m_+ \cap (\lambda_1,\lambda_2)$ is open, it follows that $m_+ \cap [\lambda_1',\lambda_2')$ is relatively open. This implies there is a λ^* such that

$$m_+ = [\lambda_1',\lambda^*), \; m_- = [\lambda^*,\lambda_2').$$

(a) If $\lambda^* = \lambda_1' + \frac{1}{2}(\lambda_2' - \lambda_1')$ then $[\lambda^*,\lambda_2') = m_-$ has length $\frac{1}{2}(\lambda_2' - \lambda_1')$. We have

$$\lambda_1' \in m_+.$$

Consequently there exists $\mu_0 \in (\lambda_1',\lambda_2]$ such that

$$\frac{\rho(\lambda_1') - \rho(\mu_0)}{\lambda_1' - \mu_0} > \zeta.$$

In particular,

$$\sup_{\mu\in[\lambda_1',\mu_0)} \frac{\rho(\mu_0) - \rho(\mu)}{\mu_0 - \mu} > \zeta.$$

This implies

$$\mu_0 \in m_- \quad \text{or} \quad \mu_0 \in [\lambda^*,\lambda_2').$$

In any case

$$|\lambda_1' - \mu_0| \geq \tfrac{1}{2}(\lambda_2' - \lambda_1')$$

so that

$$\rho(\mu_0) - \rho(\lambda_1') > \zeta\ (\mu_0 - \lambda_1') \geq \frac{5}{2}\ (\lambda_2' - \lambda_1') = \rho(\lambda_2) - \rho(\lambda_1)$$

which contradicts the monotonicity of ρ.

(b) If $\lambda^* \neq \lambda_1' + \tfrac{1}{2}(\lambda_2' - \lambda_1')$, then m_+ or m_- has length larger than $\tfrac{1}{2}(\lambda_2' - \lambda_1')$. If m_+ is the larger part then the argument is the same as above. If m_- is the larger part, then consider

$$\lambda_2^* = \lambda^* + \tfrac{1}{2}(\lambda_2' - \lambda_1') \in m_-.$$

First, we have

$$\frac{\rho(\lambda_2^*) - \rho(\mu_0)}{\lambda_2^* - \mu_0} > \zeta \quad \text{for some } \mu_0 < \lambda_2^*.$$

As in (a), we see that $\mu_0 \in [\lambda_1', \lambda^*)$. Thus

$$\lambda_2^* - \mu_0 > \tfrac{1}{2}(\lambda_2' - \lambda_1'),$$

which again leads to a contradiction.

Case 2: If m_+, m_- are not disjoint, at least one of these sets has measure larger than $\tfrac{1}{2}(\lambda_2' - \lambda_1')$. We choose finitely many disjoint left closed intervals contained in this set such that the measure m^* of the union of these intervals is again larger than $\tfrac{1}{2}(\lambda_2' - \lambda_1')$. Then as above

$$m^* > \tfrac{1}{2}(\lambda_2' - \lambda_1')$$

leads to a contradiction. □

We shall need the following approximation lemma (a generalization of Rellich's theorem).

Lemma 9.7: Let H be a separable Hilbert space with inner product $(\cdot,\cdot)$ and norm $\|\cdot\|$. Let $\{(\cdot,\cdot)_n\}_{n=1}^{\infty}$ be a sequence of inner products on H uniformly equivalent to $(\cdot,\cdot)$ and denote H equipped with this inner product $(\cdot,\cdot)_n$ by H_n. Let

$$(\cdot,\cdot)_n \to (\cdot,\cdot) \text{ uniformly as } n \to \infty.$$

Let

$$A_n : \mathcal{D}(A_n) \subset H_n \to H_n, \quad n \in \mathbb{N}, \; A : \mathcal{D}(A) \subset H \to H,$$

be self-adjoint operators satisfying

$$(A_n - i)^{-1} \underset{s}{\to} (A-i)^{-1} \text{ as } n \to \infty,$$

where $\underset{s}{\to}$ denotes strong convergence.

Let $\lambda_0 \in \mathbb{R}$ exist such that

$$N(A - \lambda_0) \subset \bigcap_{n=1}^{\infty} \mathcal{D}(A_n)$$

and let

$$a_n := \|(A_n - \lambda_0)|_{N(A - \lambda_o)}\| \to 0 \quad \text{as} \quad n \to \infty.$$

Then, for the respective spectral families $\{E_n(\lambda), \lambda \in \mathbb{R}\}$ and $\{E(\lambda), \lambda \in \mathbb{R}\}$ of A_n and A we have

$$E_n(\lambda_0 + (-\varepsilon_n^1, \varepsilon_n^2)) \underset{s}{\to} E(\{\lambda_0\}) \quad \text{as} \quad n \to \infty, \tag{9.29}$$

for any positive null sequences $\{\varepsilon_n^i\}_{n=1}^{\infty}$ satisfying the convergence condition

$$\frac{a_n}{\varepsilon_n^i} \to 0 \quad \text{as} \quad n \to \infty, \quad i = 1,2. \tag{9.30}$$

If

$$a_{n_o} = 0$$

for some $n_o \in \mathbb{N}$, then

$$E_{n_o}(\lambda_o + (-\varepsilon^1_{n_o}, \varepsilon^2_{n_o})) = E_{n_o}(\{\lambda_o\}). \tag{9.31}$$

If, in particular, $\lambda_o \notin \sigma_p(A)$ then

$$E_n(\lambda_n) \underset{s}{\rightarrow} E(\lambda_o) \quad \text{as} \quad n \rightarrow \infty, \tag{9.32}$$

for any sequence $\{\lambda_n\}_{n=1}^{\infty} \subset \mathbb{R}$, such that $\lambda_n \rightarrow \lambda_o$ as $n \rightarrow \infty$.

<u>Proof:</u> We first show that, for any $z \in \mathbb{C} \setminus \mathbb{R}$,

$$(A_n - z)^{-1} \underset{s}{\rightarrow} (A-z)^{-1} \quad \text{as} \quad n \rightarrow \infty.$$

Define

$$B_n := (A_n - z)\,(A_n - i)^{-1},$$
$$B_n := (A-z)\,(A-i)^{-1}.$$

Then

$$B_n = Id + (i-z)(A_n - i)^{-1},$$
$$B = Id + (i-z)(A-i)^{-1}.$$

Consequently

$$B_n \underset{s}{\rightarrow} B \quad \text{as} \quad n \rightarrow \infty.$$

By the equivalence of norms there exists C, c' > 0 such that, for all $f \in H$

$$C\|B_nf\|^2 \geq \|B_nf\|_n^2 = \|\int_{\mathbb{R}} \frac{(\lambda-z)}{(\lambda-i)}\, dE_n(\lambda)f\|_n^2$$

$$= \int_{\mathbb{R}} \frac{|\lambda-z|^2}{|\lambda-i|^2}\, d\|E_n(\lambda)f\|_n^2 \geq \min_{\lambda\in\mathbb{R}} \frac{(\text{Im } z)^2 + (\lambda-\text{Re } z)^2}{\lambda^2+1} \|f\|_n^2$$

$$\geq c'\|f\|.$$

Therefore, we see that

B_n^{-1} are uniformly bounded in H.

Thus, for all $f \in H$

$$\|(B_n^{-1} - B^{-1})f\| = \|B_n^{-1}(B-B_n)B^{-1}f\|$$

$$\leq C/c' \|(B-B_n)B^{-1}f\| \to 0 \quad \text{as} \quad n \to \infty.$$

Since for Im $z \neq i$ we have

$$B_n^{-1} = \text{Id} + (z-i)(A_n-z)^{-1},$$

$$B^{-1} = \text{Id} + (z-i)(A-z)^{-1},$$

it follows that, for any $z \in \mathbb{C} \setminus \mathbb{R}$

$$(A_n - z)^{-1} \underset{s}{\to} (A-z)^{-1} \quad \text{as} \quad n \to \infty.$$

Now, define

$$F_\lambda(\mu) := (\mu-\lambda-i)^{-1} + (\mu-\lambda+i)^{-1} = \frac{2(\mu-\lambda)}{1+(\mu-\lambda)^2}.$$

Then $W_\lambda := F_\lambda(A)$, $W_{\lambda,n} := F_\lambda(A_n)$ are self-adjoint operators with spectral families

$$\{E_\lambda(t) = E \circ F_\lambda^{-1}(t),\ E_{\lambda,n}(t) = E_n \circ F_{\lambda,n}^{-1}(t),\quad t \in \mathbb{R}\}$$

respectively.

In particular,

$$F_\lambda(\lambda) = 0, \; F_\lambda(\mu) \neq 0 \quad \text{for} \quad \mu \neq \lambda$$

so that

$$E_\lambda(0) = E(\lambda).$$

Similarly,

$$E_{n,\lambda}(0) = E_n(\lambda).$$

Now, define bounded, self-adjoint, positive definite operators S_n by requiring that $(u,v)_n = (u,S_n v)$ for all $u,v \in H$. By assumption we have

$$S_n \underset{s}{\rightarrow} \mathrm{Id} \quad \text{as} \quad n \rightarrow \infty.$$

From the resolvent convergence above we get

$$S_n^{1/2} \, W_{\lambda,n} \, S_n^{-1/2} \underset{s}{\rightarrow} W_\lambda \quad \text{as} \quad n \rightarrow \infty.$$

$S_n^{1/2} \, W_{\lambda,n} \, S_n^{-1/2}$ is self-adjoint in H with spectral family

$$\{S_n^{1/2} \, E_{\lambda,n}(\mu) \, S_n^{-1/2}, \; \mu \in \mathbb{R}\}.$$

According to Rellich's theorem we have, in particular, that

$$\begin{aligned} \underset{n\rightarrow\infty}{\text{s-lim}} \, S_n^{1/2} \, E_{n,\lambda}(0) \, S_n^{-1/2} &= \underset{n\rightarrow\infty}{\text{s-lim}} \, E_{n,\lambda}(0) \\ &= \underset{n\rightarrow\infty}{\text{s-lim}} \, E_n(\lambda) \\ &= E_\lambda(0) = E(\lambda). \end{aligned}$$

Note that $0 \in \sigma_p(W_\lambda)$ if $\lambda \in \sigma_p(A)$. Now, let $\{\varepsilon_n^i\}_{n=1}^\infty$, $i = 1,2$, be as stated in the theorem. For $u \in H$, $\varphi \in N(A-\lambda_o)$ we obtain

$$|((Id-E_n(\lambda_o + (-\varepsilon_n^1,\varepsilon_n^2)))u,\varphi)_n|$$

$$= |((A_n - \lambda_o)^{-1}(Id-E_n(\lambda_o + (-\varepsilon_n^1,\varepsilon_n^2)))u,(A_n - \lambda_o)\varphi)_n|$$

$$\leq \frac{a_n}{\min(\varepsilon_n^1,\varepsilon_n^2)} \|u\|_n \|\varphi\|_n.$$

If Q_n is the H_n-orthogonal projection on $N(A-\lambda_o)$ then this estimate may be rephrased as

$$|(Q_n(Id-E_n(\lambda_o + (-\varepsilon_n^1,\varepsilon_n^2)))u,v)|$$

$$\leq \frac{a_n}{\min(\varepsilon_n^1,\varepsilon_n^2)} \|u\|_n \|v\|_n, \quad u,v \in H.$$

Consequently,

$$Q_n E_n(\lambda_o + (-\varepsilon_n^1,\varepsilon_n^2)) \underset{s}{\rightarrow} E(\{\lambda_o\}) \quad \text{as} \quad n \rightarrow \infty,$$

since

$$Q_n \underset{s}{\rightarrow} E(\{\lambda_o\}) \quad \text{as} \quad n \rightarrow \infty.$$

Now,

$$Q_n = (S_n)_Q^{-1} Q S_n,$$

in the notation of Lemma 9.3, with $Q = E(\{\lambda_o\})$. Thus, here we see that

$$Q_n \underset{s}{\rightarrow} Q \quad \text{as} \quad n \rightarrow \infty.$$

In fact,

$$((S_n)_Q^{-1} u, v)_n = (u,v) \text{ for } u,v \in N(A-\lambda_o).$$

Moreover, we claim that

$$E_n(\lambda_o + (-\varepsilon_n^1, \varepsilon_n^2)) \underset{s}{\rightarrow} E(\{\lambda_o\}) \quad \text{as} \quad n \rightarrow \infty.$$

This can be shown by contradiction. Assume that there exists $\varepsilon > 0$ such that

$$\|(\mathrm{Id} - Q_{n_k})E_{n_k}(\lambda_o + (-\varepsilon_{n_k}^1, \varepsilon_{n_k}^2))f\|^2 \geq \varepsilon$$

for some subsequence $\{Q_{n_k}\}_k$ of $\{Q_n\}_n$. Then

$$\lim_{k\to\infty} \|E_{n_k}(\lambda_o + (-\varepsilon_{n_k}^1, \varepsilon_{n_k}^2))f\|^2 \geq \|E(\{\lambda_o\})f\|^2 + \varepsilon.$$

On the other hand, for any $\varepsilon^1, \varepsilon^2$, we have

$$\varepsilon_{n_k}^1 < \varepsilon^1, \ \varepsilon_{n_k}^2 < \varepsilon^2,$$

for all k larger than some $k_o \in \mathbb{N}$. Thus

$$\|E_{n_k}(\lambda_o + (-\varepsilon^1, \varepsilon^2))f\|^2 \geq \|E_{n_k}(\lambda_o + (-\varepsilon_{n_k}^1, \varepsilon_{n_k}^2))f\|^2$$

for $k \geq k_o$, and so

$$\lim_{k\to\infty} \|E_{n_k}(\lambda_o + (-\varepsilon^1, \varepsilon^2))f\|^2 \geq \|E(\{\lambda_o\})f\|^2 + \varepsilon.$$

For $\lambda_o + \varepsilon^2$, $\lambda_o - \varepsilon^1 \notin \sigma_p(A)$, we have, according to the first part of the proof,

$$\|E(\lambda_o + (-\varepsilon^1, \varepsilon^2))f\| \geq \|E(\{\lambda_o\})f\| + \varepsilon$$

for arbitrary small such $\varepsilon^1, \varepsilon^2$. This contradicts the right continuity of the spectral function E. □

<u>Lemma 9.8:</u> Let Z_k, $k = 1,\ldots,n$, be orthogonal projections in a separable Hilbert space H satisfying

$$Z_k Z_j = \frac{1}{\gamma_k - \gamma_j} Z_k(M_k^* N_j - N_k^* M_j)Z_j, \quad k,j = 1,\ldots,n, \ k \neq j, \tag{9.33}$$

where $\{\gamma_k\}_{k=0}^n \subset \mathbb{R}$, $\gamma_{k-1} < \gamma_k$, and M_k, N_k bounded, $k = 1,\dots,n$. Moreover, let

$$\|M_k Z_k\| \leq c\ |\gamma_k - \gamma_{k-1}|, \quad \|N_k Z_k\| \leq c', \quad k = 1,\dots,n,$$

for some $c, c' > 0$. Then for the operator

$$Z := \sum_{k=1}^{n} Z_k \ : \ H \longrightarrow H$$

we have

$$0 \leq Z \leq (1 + 4\ c'c\ \frac{b}{a}\ (\pi+b+1)) \tag{9.34}$$

where

$$a = \min_k |\gamma_k - \gamma_{k-1}|, \quad b = \max_k |\gamma_k - \gamma_{k-1}|.$$

The proof relies on the following lemma.

<u>Lemma 9.9:</u> Let $\{\alpha_i\}_{i=0}^n \subset \mathbb{R}$ be such that

$$0 < a \leq \alpha_k - \alpha_{k-1} \leq b, \quad k = 1,\dots,n$$

for some $a, b \in \mathbb{R}$. Then

$$\Big| \sum_{k,j=1}^{n} \frac{x_k\ y_j}{\alpha_k - \alpha_j} \Big|^2 \leq 4 \left(\frac{\pi+b+1}{a}\right)^2 \sum_{j=1}^{n} |x_j|^2 \sum_{k=1}^{n} |y_k|^2, \tag{9.35}$$

for all n-tuples $(x_k)_k, (y_j)_j \in \mathbb{R}^n$,

<u>Proof:</u> First, define step functions α, β by

$$\alpha(x) := \frac{x_k}{\alpha_k - \alpha_{k-1}} \quad \text{for} \quad x \in (\alpha_{k-1}, \alpha_k], \quad k = 1,\dots,n,$$

$$\beta(y) := \frac{y_k}{\alpha_k - \alpha_{k-1}} \quad \text{for} \quad y \in (\alpha_{k-1}, \alpha_k], \quad k = 1,\dots,n.$$

The idea of the proof is to regard the expression

$$H(y) := (\pi^{-1} \sum_{\substack{j=1 \\ j\neq k}}^{n} \frac{y_j}{\alpha_k - \alpha_j})_{k=1}^{n} \in \mathbb{C}^n$$

as a discrete Hilbert transform and to compare it with the (continuous) Hilbert transform

$$HT(f)(x) = \pi^{-1} \text{ P.V.} \int_{\mathbb{R}} \frac{f(y)}{x-y} \, dy \tag{9.36}$$

in order to obtain the desired estimate. Since HT is unitary in $L_2(\mathbb{R})$ and its inverse is -HT, we have

$$(\alpha, HT(\beta))_{L_2(\mathbb{R})} = -(HT(\alpha), \beta)_{L_2(\mathbb{R})}$$

and (9.37)

$$|(\alpha, HT(\beta))_{L_2(\mathbb{R})}| \le \|\alpha\|_{L_2(\mathbb{R})} \|\beta\|_{L_2(\mathbb{R})} .$$

For sake of simplicity of notation we shall drop the index $L_2(\mathbb{R})$ on the norm and inner product of $L_2(\mathbb{R})$ for the rest of the proof. We estimate

$$|\pi^{-1} \sum_{j\neq k} \frac{\overline{x}_k \, y_j}{\alpha_k - \alpha_j} - (\alpha, HT(\beta))|$$

$$\le \sum_{|j-k|>1} |x_k| \, |y_j| \left| \frac{1}{\pi(\alpha_k - \alpha_j)} - \frac{(\chi_{[\alpha_{k-1},\alpha_k]}, HT(\chi_{[\alpha_j,\alpha_{j-1}]}))}{(\alpha_k - \alpha_{k-1})(\alpha_j - \alpha_{j-1})} \right| \tag{9.38}$$

$$+ (a\pi)^{-1} (1+\pi) \sum_{|j-k|=1} |x_k| \, |y_j| .$$

For the first term on the right-hand side of this estimate we have

$$\sum_{|j-k|>1} |x_k| \, |y_j| \left| \frac{1}{\pi(\alpha_k - \alpha_j)} - \frac{(\chi_{[\alpha_{k-1},\alpha_k]}, HT(\chi_{[\alpha_j,\alpha_{j-1}]}))}{(\alpha_k - \alpha_{k-1})(\alpha_j - \alpha_{j-1})} \right|$$

$$\le 2 \sum_{k>j+1} |x_k| \, |y_j| \left| \frac{1}{\pi(\alpha_k - \alpha_j)} - \frac{1}{\pi(\alpha_{k-1} - \alpha_j)} \right|$$

$$\leq \frac{2b}{\pi} \sum_{k>j+1} \frac{1}{(\alpha_k-\alpha_j)(\alpha_{k-1}-\alpha_j)} |x_k| \, |y_j|$$

$$\leq \frac{2b}{\pi a} \sum_{k>j+1} \frac{|x_k| \, |y_j|}{(k-j)(k-j-1)} \quad .$$

This in turn may be estimated further using the following result of J. Schur, [8]:

Let

$$\sum_{q=1}^{\infty} |a_{pq}| \leq c_1 \text{ and } \sum_{p=1}^{\infty} |a_{pq}| \leq c_2.$$

Then

$$|\sum_{p,q} |a_{pq}| \, x_p \, y_q| \leq (c_1 c_2)^{1/2} \, (\sum_p |x_p|^2)^{1/2} \, (\sum_q |y_q|^2)^{1/2}.$$

Since

$$\sum_{k=1}^{\infty} \frac{1}{k(k+1)} \leq 1$$

we obtain

$$\begin{aligned} &\sum_{|j-k|>1} |x_k| \, |y_j| \left| \frac{1}{\pi(\alpha_k-\alpha_j)} - \frac{(\chi_{[\alpha_{k-1},\alpha_k]}, \, HT(\chi_{[\alpha_j,\alpha_{j-1}]}))}{(\alpha_k-\alpha_{k-1})(\alpha_j-\alpha_{j-1})} \right| \\ &\leq \frac{2b}{\pi a} \sum_{k>j+1} \frac{|x_k| \, |y_j|}{(k-j)(k-j-1)} \leq \frac{2b}{\pi a} \, (\sum_k |x_k|^2)^{1/2} \, (\sum_j |y_j|^2)^{1/2}. \end{aligned} \tag{9.39}$$

Finally, from (9.38) and (9.39) we get

$$\begin{aligned} \left| \pi^{-1} \sum_{j \neq k} \frac{\overline{x}_k \, y_j}{\alpha_k-\alpha_j} \right| &\leq \left| \pi^{-1} \sum_{j \neq k} \frac{\overline{x}_k \, y_j}{\alpha_k-\alpha_j} - (\alpha, \, HT(\beta)) \right| + |(\alpha, \, HT(\beta))| \\ &\leq \frac{2(b+1+\pi)}{\pi a} \, (\sum_k |x_k|^2)^{1/2} \, (\sum_j |y_j|^2)^{1/2}. \qquad \square \end{aligned}$$

Proof of Lemma 9.8: A straightforward calculation yields

$$
\begin{aligned}
\Big|\sum_{k\neq j}(Z_kf,Z_jf)\Big|^2 &= \Big|\sum_{k\neq j}(f,Z_kZ_jf)\Big|^2 \\
&= \Big|\sum_{k\neq j}\frac{1}{\gamma_k-\gamma_j}(Z_kf,(M_k^*N_j-N_k^*M_j)Z_jf)\Big|^2 \\
&= \Big|\sum_{k\neq j}\frac{1}{\gamma_k-\gamma_j}(M_kZ_kf,N_jZ_jf)-\sum_{k\neq j}\frac{1}{\gamma_k-\gamma}(N_kZ_kf,M_jZ_jf)\Big|^2 \\
&\le \Big|\sum_{k\neq j}\frac{1}{|\gamma_k-\gamma_j|}\|M_kZ_kf\|\|N_jZ_jf\|+\sum_{k\neq j}\frac{1}{|\gamma_k-\gamma_j|}\|N_kZ_kf\|\|M_jZ_jf\|\Big|^2 \\
&\le 16\left(\frac{\pi+b+1}{a}\right)^2\sum_k\|M_kZ_kf\|^2\sum_j\|N_jZ_jf\|^2, \quad \text{using Lemma 9.8,} \\
&\le 16\, c'^2c^2b^2\left(\frac{\pi+b+1}{a}\right)^2\Big(\sum_k\|Z_kf\|^2\Big)^2.
\end{aligned}
$$

Thus

$$
\begin{aligned}
\|Zf\|^2 = \Big\|\sum_k Z_kf\Big\|^2 &= \sum_k\|Z_kf\|^2+\sum_{k\neq j}(Z_kf,Z_jf) \\
&\le \Big(1+4c'c\,\frac{b}{a}\,(\pi+b+1)\Big)\sum_k\|Z_kf\|^2 \\
&\le \Big(1+4c'c\,\frac{b}{a}\,(\pi+b+1)\Big)(Zf,f)
\end{aligned}
$$

This gives the required bound. □

10 A two-parameter spectral representation

In order to obtain the desired spectral representation we have to impose one more restrictive assumption on the two-parameter problem. Nevertheless, many applications are still covered by this approach. We assume that one of the operators, say T_1, has discrete spectrum. It is an open question under what conditions a similar result holds for the case that both operators have partly continuous spectrum. Thus, for the following we assume:

$$T_1 : \mathcal{D}(T_1) \subset H_1 \longrightarrow H_1 \text{ has purely discrete spectrum.} \tag{10.1}$$

Under this general assumption, consider the corresponding set of eigencurves in $\mathbb{R}^2$ of the operator $(T-V \circ \Lambda)_1 : \mathcal{D}(T_1) \subset H_1 \longrightarrow H_1$, $\Lambda \in \mathbb{R}^2$, i.e.

$$\{\Lambda \in \mathbb{R}^2 \,|\, N((T-V \circ \Lambda)_1) \neq \{0\}\}. \tag{10.2}$$

It is known that this set is the union of a countable number of analytic curves $\{C_\nu^1\}_{\nu=1}^\infty$ and that the set of intersection points of these curves is discrete (See [7]). Moreover,

$$\left.\begin{array}{l} \dim N((T-V \circ \Lambda)_1) < \infty \\ \quad R((T-V \circ \Lambda)_1) \text{ closed} \end{array}\right\} \quad \forall \Lambda \in \mathbb{R}^2. \tag{10.3}$$

For $\Lambda \in C_\nu^1$ there exists an orthonormal system $S_\nu(\Lambda) = \{\varphi_{\nu,k}^1(\Lambda)\}_{k=1}^{n_\nu} \subset H_1$, depending analytically on the curve parameter. For any $\Lambda \in \mathbb{R}^2$

$$\bigcup_{\nu:\Lambda\in C_\nu^1} S_\nu(\Lambda) \quad \text{spans} \quad N((T-V \circ \Lambda)_1)$$

where the union takes account of points where the curves intersect. Note that these are properties of the original operators in H_1 not of the 'tensorized' operators in H. Though we did agree upon not distinguishing these by notation it should be kept in mind that here this difference is nonetheless important.

For the slope of C^1_ν we have

$$\frac{d\lambda_1}{d\lambda_2}(\Lambda) = -\frac{\|\varphi^1_{\nu,k}(\Lambda)\|^2_{12}}{\|\varphi^1_{\nu,k}(\Lambda)\|^2_{11}}, \quad k = 1,\dots,n_\nu. \tag{10.4}$$

For any fixed ν, let b be a left open, right closed connected piece of C^1_ν (i.e. an arc) with end points Λ^-, Λ^+. Moreover, let Λ_0 denote the midpoint of the line segment s_0 connecting Λ^+ and Λ^-, and let ϑ_0 be the angle between this line segment and the λ_2-axis, $-\frac{\pi}{2} < \vartheta_0 < 0$. Finally, let β_0 denote the distance from Λ_0 to Λ^+ and Λ^- (see Figure 1).

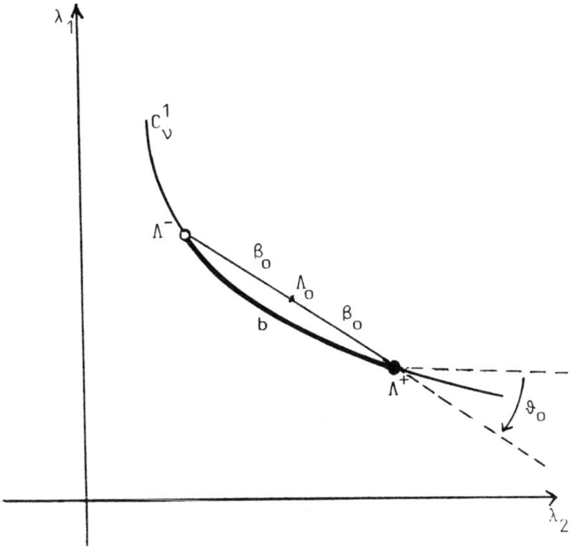

Fig. 1

For $\Lambda^* \in b$ we define

$$G_0(b) := E_{\vartheta_0,\Lambda^*,11}(\{0\})\, E_{\vartheta_0,\Lambda_0,22}((-\beta_0,\beta_0]),$$

$$G(b) := \bigoplus_{i=1}^{2} G_0(b).$$

Renormalization of $G_0(b)$, $G(b)$ with respect to

$$V(\vartheta_o) := \Delta_o (V \circ \Theta(\vartheta_o))^{-1}_{11} (V \circ \Theta(\vartheta_o))^{-1}_{22}$$

where $\Theta(\cdot)$ is the rotation matrix introduced in section 8, leads, according to Lemma 9.3, to new projections (having the same range, but which are now orthogonal in the sense of H) that will be denoted by $F_o(b)$, $F(b)$, respectively, and it is clear that

$$F(b) = F_o(b) \oplus F_o(b). \tag{10.5}$$

We shall see that $F(b)$ is a good approximation of $\Pi(b)$.

Let us assume the following situation: $b \subset \hat{b} \subset C^1_\nu$, where $\hat{b}$ is closed and connected and does not contain intersection points of C^1_ν with other eigencurves. Moreover we assume $\beta_o < \hat{\beta}$, for some $\hat{\beta} < 1$ to be fixed later. Furthermore, let $\hat{I} = \hat{I}_1 \times \hat{I}_2$ be a fixed rectangle with sides $\hat{I}_1, \hat{I}_2$ containing $\hat{b}$.

Definition 10.1: $\Lambda \in \mathbb{R}^2$ is called a simultaneous eigenvalue of Γ if

$$\Pi(\{\Lambda\}) \neq 0.$$

In the following it will again be important to employ the rotation matrix $\Theta(\vartheta)$. Note that

$$\begin{aligned} T - V \circ (\Lambda_o + \alpha) &= (T - V \circ \Lambda_o) - V \circ \alpha \\ &= (T - V \circ \Lambda_o) - (V \circ \Theta(\vartheta)) \circ (\hat{\Theta}(\vartheta) \circ \alpha) \\ &= (T - V \circ \Lambda_o) - (V \circ \Theta(\vartheta)) \circ \tilde{\alpha}, \end{aligned}$$

where

$$\tilde{\alpha} := \hat{\Theta}(\vartheta) \circ \alpha \text{ and } \hat{\Theta}(\vartheta) \text{ is the cofactor matrix of } \Theta(\vartheta)$$

is again a two-parameter system with parameters $\alpha \in \mathbb{R}^2$. Let $\Gamma(\vartheta, \Lambda_o)$ be the operator constructed for this system analogously to Γ for the original system. Then by a simple calculation we see that

$$\Gamma(\vartheta,\Lambda_o) = \hat{\Theta}(\vartheta) \circ (\Gamma - \Lambda_o).$$

In section 8 we used a rotation by $\vartheta = -\frac{\pi}{4}$. Here we shall make use of the ability to adapt the angle ϑ more closely to the eigencurve. $\Gamma(\vartheta,\Lambda_o)$ has, analogous to Γ, two components $\Gamma_i(\vartheta,\Lambda_o)$, $i = 1,2$:

$$\Gamma(\vartheta,\Lambda_o) = \Gamma_1(\vartheta,\Lambda_o) \oplus \Gamma_2(\vartheta,\Lambda_o).$$

We have the following approximation lemma:

Lemma 10.2: Suppose that neither Λ^+ nor Λ^- are simultaneous eigenvalues of $\overline{\Gamma}$. Then for $\varkappa \in (0, \frac{1}{2}]$, $\beta_o < \hat{\beta}$, we have

$$\begin{aligned} &|[g,(\Pi(b) - F(b))\Pi(\hat{I})f]| \\ &\leq \beta_o^{3/5}\, c(\varkappa,\hat{I})\, \{|||g|||\ |||W_b^{\varkappa}\, \Pi(b)f||| + |||F(b)g|||\ (|||f||| + |||W_b^{\varkappa} f|||)\} \end{aligned} \tag{10.6}$$

for all $g \in H$, $f \in \mathcal{D}(W_b^{\varkappa})$ where

$$\begin{aligned} W_{o,b} &:= \beta_o^2\, |\Gamma_2(\vartheta_o,\Lambda^+)\ \Gamma_2(\vartheta_o,\Lambda^-)|^{-1} \\ W_b &:= W_{o,b} \oplus W_{o,b}\,. \end{aligned} \tag{10.7}$$

Proof: The proof uses Corollary 7.2 and applies Lemma 9.1 and Lemma 9.2 several times.

Consider the line through a point $\tilde{\Lambda} \in \mathbb{R}^2$ with angle $\tilde{\vartheta}$ to the λ_2-axis. Let α_m, α^m be the infimum and supremum of the distance between a given arc b and this line, and β_m, β^m the minimal and maximal distance of the projection of b on this line to $\tilde{\Lambda}$. Finally, denote by δ_m, δ^m the minimal and maximal angles between directions in $b - \tilde{\Lambda}$ and the direction of the line (see Figure 2). Then

$$\alpha_m |||E(b)u|||^{\otimes} \leq |||\overline{\Gamma}_1(\tilde{\vartheta}, \tilde{\Lambda})\ E(b)u|||^{\otimes} \leq \alpha^m |||E(b)u|||^{\otimes},$$

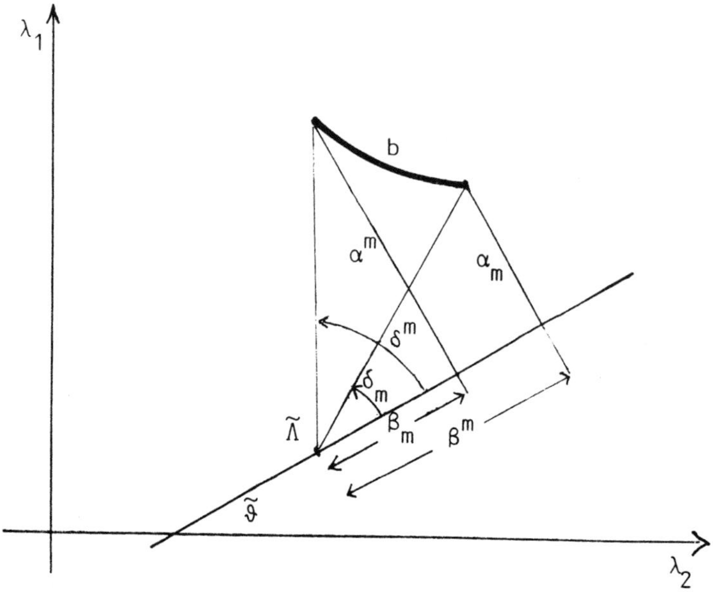

Fig. 2

since, by the spectral theorem,

$$\left\{\int_{\mu \in b} |(\mu_1 - \tilde{\lambda}_1)\cos\tilde{\vartheta} - (\mu_2 - \tilde{\lambda}_2)\sin\tilde{\vartheta}|^2 \; d|||E(\mu)u|||^2\right\}^{1/2}$$

$$= |||\overline{\Gamma}_1(\tilde{\vartheta},\tilde{\Lambda}) \; E(b)u|||^{\otimes}.$$

Similarly

$$\beta_m |||E(b)u|||^{\otimes} \leq |||\overline{\Gamma}_2(\tilde{\vartheta},\tilde{\Lambda}) \; E(b)u|||^{\otimes} \leq \beta^m |||E(b)u|||^{\otimes} \tag{10.8}$$

and

$$\tan \delta_m \; |||E(b)u|||^{\otimes} \leq |||\overline{\Gamma}_1(\tilde{\vartheta},\tilde{\Lambda}) \; \overline{\Gamma}_2(\tilde{\vartheta},\tilde{\Lambda})^{-1} \; E(b)u|||^{\otimes} \leq \tan \delta^m |||E(b)u|||^{\otimes}. \tag{10.9}$$

For the latter note that

$$\tan \delta_m \le \frac{|(\mu_1 - \tilde{\lambda}_1)\cos\tilde{\vartheta} - (\mu_2 - \tilde{\lambda}_2)\sin\tilde{\vartheta}|}{|(\mu_1 - \tilde{\lambda}_1)\sin\tilde{\vartheta} + (\mu_2 - \tilde{\lambda}_2)\cos\tilde{\vartheta}|} \le \tan \delta^m$$

for all $u \in H^{\otimes}$ provided $\delta^m < \frac{\pi}{2}$.

We introduce new coordinates (α,β) around Λ_0 by setting

$$\Lambda = \Lambda_0 + \Theta(\vartheta_0)\begin{pmatrix}\alpha\\ \beta\end{pmatrix}.$$

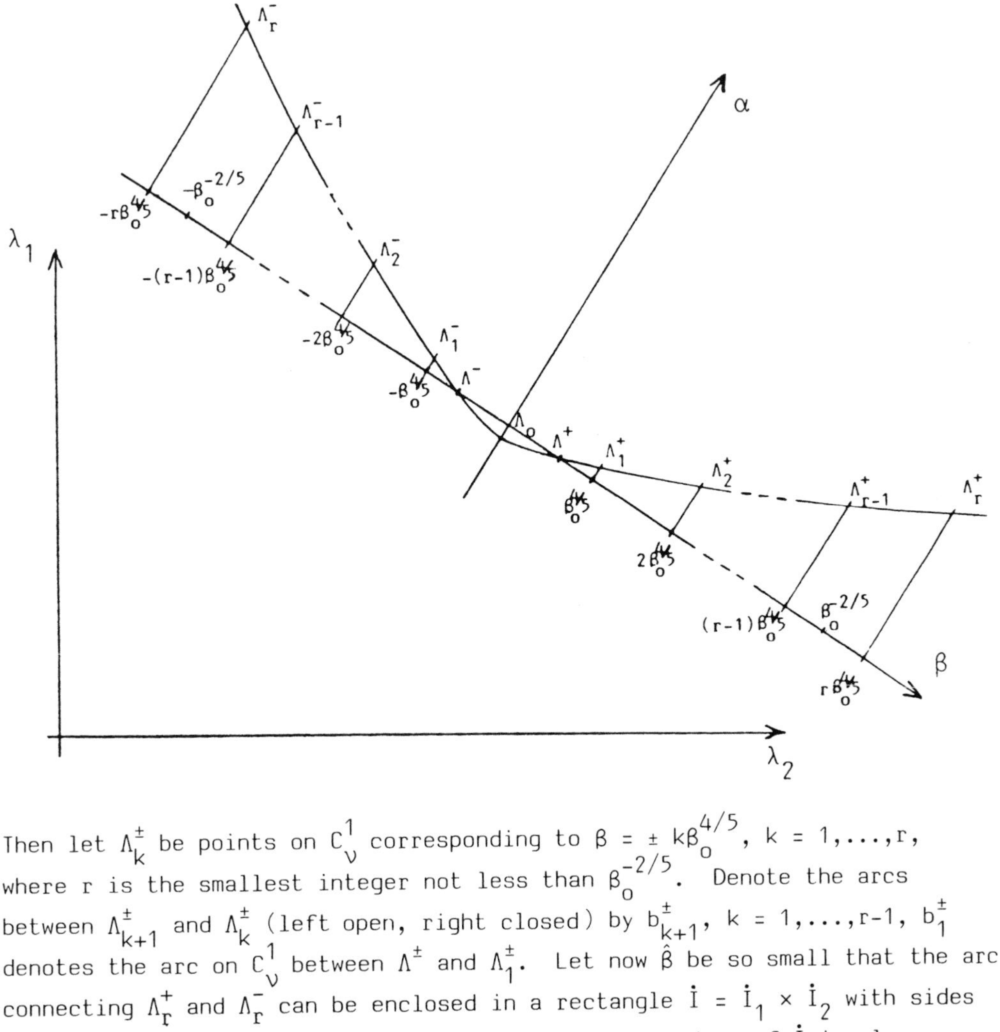

Then let $\Lambda_k^{\pm}$ be points on C_ν^1 corresponding to $\beta = \pm k\beta_0^{4/5}$, $k = 1,\dots,r$, where r is the smallest integer not less than $\beta_0^{-2/5}$. Denote the arcs between $\Lambda_{k+1}^{\pm}$ and $\Lambda_k^{\pm}$ (left open, right closed) by $b_{k+1}^{\pm}$, $k = 1,\dots,r-1$, $b_1^{\pm}$ denotes the arc on C_ν^1 between $\Lambda^{\pm}$ and $\Lambda_1^{\pm}$. Let now $\hat{\beta}$ be so small that the arc connecting Λ_r^+ and Λ_r^- can be enclosed in a rectangle $\dot{I} = \dot{I}_1 \times \dot{I}_2$ with sides $\dot{I}_1, \dot{I}_2$ such that the distance between Λ_0 and the boundary of $\dot{I}$ is always larger than $c_1\,\beta_0^{2/5}$ for some $c_1 > 0$ and $\dot{I}$ has no common points with eigencurves other than C_ν^1. In this situation we have

$$
\left.\begin{aligned}
&(E(b) - F_o(b))E(\hat{I}) = (Id-F_o(b))E(b) - F_o(b)(E(\hat{I}) - E(\dot{I})) \\
&- \sum_{k=2}^{r} F_o(b)(E(b_k^+) + E(b_k^-)) - F_o(b)(E(b_1^+) + E(b_1^-)).
\end{aligned}\right\} \qquad (10.10)
$$

Equation (10.10) follows immediately from

$$
\left.\begin{aligned}
&(E(b) - F_o(b))E(\hat{I}) = (Id-F_o(b))E(b) - F_o(b)(E(\hat{I}) - E(b)) \\
&E(\hat{I}) - E(b) = E(\hat{I}) - E(\dot{I}) + E(\dot{I}) - E(b) \\
&E(\dot{I}) - E(b) = \sum_{k=2}^{r} (E(b_k^+) + E(b_k^-)) + (E(B_1^+) + E(b_1^-)).
\end{aligned}\right\} \qquad (10.11)
$$

In order to achieve the result we shall estimate each of the four terms on the right-hand side of (10.10) separately.

(1) Estimate of $(Id-F_o(b))E(b)$:

We have, by the construction of $F_o(b)$, equality of ranges, i.e.

$$
R(G_o(b)) = R(F_o(b)),
$$

so that

$$
(Id-F_o(b))\, E(b) = (Id-F_o(b))(Id-G_o(b))\, E(b).
$$

Thus, we obtain the estimate

$$
\begin{aligned}
&|||(Id-F_o(b))\, E(b)f|||^{\otimes} \\
&\leq \sqrt{2}\, \|(Id-G_o(b))E(b)f\|^{\otimes} \\
&\leq \sqrt{2}\, \|(Id-E_{\vartheta_o,\Lambda^*,11}(\{0\}))\, E(b)f\|^{\otimes} \\
&+ \sqrt{2}\, \|E_{\vartheta_o,\Lambda^*,11}(\{0\})(Id-E_{\vartheta_o,\Lambda_o,22}((-\beta_o,\beta_o]))E(b)f\|^{\otimes} \\
&\leq \sqrt{2}\, \|(Id-E_{\vartheta_o,\Lambda^*,11}(\{0\}))\, E(b)f\|^{\otimes} \\
&+ \sqrt{2}\, \|(Id-E_{\vartheta_o,\Lambda_o,22}((-\beta_o,\beta_o]))\, E(b)f\|^{\otimes}.
\end{aligned}
$$

The two terms on the right-hand side will be estimated separately.

(a) To estimate the first term we apply Corollary 7.2 to the rotated operators to obtain

$$(V \circ \Theta(\vartheta))_{12}^{-1} (T-V \circ \Lambda^*)_1 = (V \circ \Theta(\vartheta))_{12}^{-1}(V \circ \Theta(\vartheta))_{11} \overline{\Gamma}_1(\vartheta,\Lambda^*)+ \overline{\Gamma}_2(\vartheta,\Lambda^*)$$

on $R(E(b))$.

For $\vartheta = (\frac{\pi}{2} + \vartheta_0) =: \vartheta^*$ we have

$$(V \circ \Theta(\vartheta_0))_{11} = (V \circ \Theta(\vartheta^*))_{12}$$

and

$$(V \circ \Theta(\vartheta_0))_{12} = -(V \circ \Theta(\vartheta^*))_{11}.$$

Consequently,

$$\begin{aligned} T_{\vartheta_0,\Lambda^*,11} &:= (V \circ \Theta(\vartheta_0))_{11}^{-1} (T-V \circ \Lambda^*)_1 \\ &= \overline{\Gamma}_2(\vartheta^*,\Lambda^*) - (V \circ \Theta(\vartheta_0))_{11}^{-1} (V \circ \Theta(\vartheta_0))_{12} \overline{\Gamma}_1(\vartheta^*,\Lambda^*) \end{aligned}$$

on $R(E(b))$. Since $T_{\vartheta_0,\Lambda^*,11}$ is boundedly invertible on $R(Id-E_{\vartheta_0,\Lambda^*,11})$, this gives

$$\begin{aligned} &(Id-E_{\vartheta_0,\Lambda^*,11}(\{0\}))\, E(b)f \\ &= (T_{\vartheta_0,\Lambda^*,11})^{-1}(Id-E_{\vartheta_0,\Lambda^*,11}(\{0\})\{\overline{\Gamma}_2(\vartheta^*,\Lambda^*) \\ &\quad - (V \circ \Theta(\vartheta_0))_{11}^{-1} (V \circ \Theta(\vartheta_0))_{12} \overline{\Gamma}_1(\vartheta^*,\Lambda^*)\}\, E(b)f. \end{aligned}$$

By applying the spectral representation of $\overline{\Gamma}$ we may estimate this as follows:

$$\begin{aligned} &\|(Id-E_{\vartheta_0,\Lambda^*,11}(\{0\}))\, E(b)f\|^{\otimes} \\ &\le c_2\, \|\{\overline{\Gamma}_2(\vartheta^*,\Lambda^*) - (V \circ \Theta(\vartheta_0))_{11}^{-1} (V \circ \Theta(\vartheta_0))_{12} \overline{\Gamma}_1(\vartheta^*,\Lambda^*)\}E(b)f\|^{\otimes} \qquad (10.12) \\ &\le c_3\, \beta_o\, |||E(b)f|||^{\otimes}. \end{aligned}$$

(b) In order to handle the second term we observe that, according to Corollary 7.2 (using a similar argument as in (a))

$$X := (Id-E_{\vartheta_0,\Lambda_0,22}((-\beta_0,\beta_0])) \, E(b) \, W_{o,b}$$

satisfies

$$X - (T_{\vartheta_0,\Lambda_0,22})^{-1} \, X \, \overline{\Gamma}_2(\vartheta_0,\Lambda_0) \, E(b)$$

$$= - (T_{\vartheta_0,\Lambda_0,22})^{-1}(Id-E_{\vartheta_0,\Lambda_0,22}((-\beta_0,\beta_0]))\cdot \tag{10.13}$$

$$\cdot(V \circ \Theta(\vartheta_o))^{-1}_{22} \, (V \circ \Theta(\vartheta_0))_{21} \, \overline{\Gamma}_1(\vartheta_0,\Lambda_0) \, W_{o,b} \, E(b).$$

The assumptions of Lemma 9.2 are satisfied. In particular

$$\|(T_{\vartheta_0,\Lambda_0,22})^{-1}(Id-E_{\vartheta_0,\Lambda_0,22}((-\beta_0,\beta_0]))\|^{\otimes} \leq \beta_0^{-1},$$

$$|||\overline{\Gamma}_2(\vartheta_0,\Lambda_0) \, E(b)|||^{\otimes} \leq \beta_0, \tag{10.14}$$

$$|||\overline{\Gamma}_1(\vartheta_0,\Lambda_0) \, W_{o,b} \, E(b)|||^{\otimes} \leq c_4 \, \beta_0^2.$$

The latter estimate shows that the right side of (10.13) is bounded. With

$$W = W_{o,b} \, E(b), \quad \varkappa_o = 1,$$

estimate (9.9) is satisfied. Thus

$$\|Xv\|^{\otimes} \leq c_5 \, \beta_0 \, c(\varkappa) \, |||W^{1+\varkappa}_{o,b} \, E(b)v|||^{\otimes} \text{ for } v \in \mathcal{D}(W^{1+\varkappa}),$$

or

$$\|(Id-E_{\vartheta_0,\Lambda_0,22}((-\beta_0,\beta_0])) \, E(b)f\|^{\otimes} \leq c_5 \, \beta_0 c(\varkappa) \, |||W^{\varkappa}_{o,b} E(b)f|||^{\otimes} \tag{10.15}$$

for all $f \in \mathcal{D}(W^{\varkappa})$. Estimates (10.12) and (10.15) together show that

$$|||(Id-F_0(b)) \, E(b)f|||^{\otimes} \leq c_6 \, \beta_0 c(\varkappa) \, |||W^{\varkappa}_{o,b} \, E(b)f|||^{\otimes}$$

for all $f \in \mathcal{D}(W^{\varkappa}_{o,b})$. Note that, by (10.8) and (10.9)

$$|||E(b)f|||^{\otimes} \leq c_7\ \beta_o^2\ |||(\overline{\Gamma}_2(\vartheta_o,\Lambda^+)\ \overline{\Gamma}_2(\vartheta_o,\Lambda^-)^{-1}\ E(b)f|||^{\otimes}$$

$$\leq c_7\ |||W_{o,b}\ E(b)f|||^{\otimes}$$

for all $f \in H^{\otimes} \cap \mathcal{D}(W_{o,b})$.

(2) Estimate of $F_o(b)\ (E(\hat{I}) - E(\dot{I}))$:

By definition of E we have

$$E(\dot{I}) = \Pi_1(\dot{I}_1)\Pi_2(\dot{I}_2)$$

so that

$$F_o(b)(E(\hat{I}) - E(\dot{I}))$$

$$= F_o(b)(Id-\Pi_2(\dot{I}_2))E(\hat{I}) + F_o(b)(Id-\Pi_1(\dot{I}_1))\ \Pi_2(\dot{I}_2)E(\hat{I}).$$

Since

$$(\Gamma-\Lambda)\Pi(\hat{I}) = \Delta_o^{-1}\ \hat{V} \circ (T-V \circ \Lambda)\Pi(\hat{I})$$

we get by the definition of $F_o(b)$

$$F_o(b)(Id-\Pi_2(\dot{I}_2))\ E(\hat{I})$$

$$= (V(\vartheta_o)_{G_o(b)})^{-1}G_o(b)V(\vartheta_o)(\Delta_o^{-1}\ \hat{V} \circ (T-V \circ \Lambda))_2(\Gamma_2-\lambda_{o,2})^{-1}(Id-\Pi_2(\dot{I}_2))E(\hat{I})$$

$$= (V(\vartheta_o)_{G_o(b)})^{-1}G_o(b)\ L(\Gamma_2-\lambda_{o,2})^{-1}(Id-\Pi_2(\dot{I}_2))E(\hat{I}), \tag{10.16}$$

where

$$L := -V_{21}\,(V \circ \Theta(\vartheta_o))_{11}^{-1}\,(T-V \circ \Lambda_o)_1\,(V \circ \Theta(\vartheta_o))_{22}^{-1}$$

$$+ V_{11}\,(V \circ \Theta(\vartheta_o))_{22}^{-1}\,(T-V \circ \Lambda_o)_2\,(V \circ \Theta(\vartheta_o))_{11}^{-1}$$

$$= -V_{21}\,T_{\vartheta_o,\Lambda_o,11}(V \circ \Theta(\vartheta_o))_{22}^{-1} + V_{11}\,T_{\vartheta_o,\Lambda_o,22}(V \circ \Theta(\vartheta_o))_{11}^{-1}.$$

First observe that

$$|||F_o(b)(\mathrm{Id}-\Pi_2(\dot{I}_2))E(\hat{I})f|||^{\otimes}$$
$$\leq \tilde{c}\,\|G_o(b)\Delta_o(V \circ \Theta(\vartheta_o))_{11}^{-1}(V \circ \Theta(\vartheta_o))_{22}^{-1}(\mathrm{Id}-\Pi_2(\dot{I}_2))\,E(\hat{I})f\|^{\otimes}. \tag{10.17}$$

Here we made use of the fact that, uniformly for $b \subset \hat{b}$,

$$|||(V(\vartheta_o)_{G_o(b)})^{-1}G_o(b)f|||^{\otimes} \leq c\;|||G_o(b)f|||^{\otimes} \tag{10.18}$$

for some constant $c > 0$ and all $f \in H^{\otimes}$. This can be seen as follows. Since by assumption the spectrum of T_1 is discrete, we have (compare the proof of Lemma 6.3)

$$\begin{aligned}\|G_o(b)f\|^{\otimes} &\leq c\,(G_o(b)f,\,V_{11}\,V_{12}\,G_o(b)f)^{1/2} \\ &\leq c(G_o(b)f,\,\Delta_o\,G_o(b)f)^{1/2} \\ &= c|||G_o(b)f|||^{\otimes}\quad \forall f \in H^{\otimes},\end{aligned} \tag{10.19}$$

where the constant is uniform with respect to $b \subset \hat{b}$. According to Lemma 9.3 we have

$$[u,(V(\vartheta_o)_{G_o(b)})^{-1}u]^{\otimes} \leq c(u,u)^{\otimes}, \tag{10.20}$$

for all $u \in R(G_o(b))$. Now (10.19) and (10.20) yield (10.18).

Using (10.16) we obtain from (10.17) (applying estimate (7.10) in the proof of Theorem 7.1)

$$|||F_o(b)(Id-\Pi_2(\dot{I}_2))E(\hat{I})f|||^{\otimes}$$

$$\leq c_8\ \beta_o\ |||(\Gamma_2-\lambda_{o,2})^{-1}(Id-\Pi_2(\dot{I}_2))\ E(\hat{I})f|||^{\otimes}.$$

To see this note that

$$\|G_o(b)\ T_{\vartheta_o,\Lambda_o,ii}\| \leq c'\ \beta_o, \quad i = 1,2.$$

Since Λ_o has distance greater than $c_1\ \beta_o^{2/5}$ from the boundary of $\dot{I}$ (by assumption), we obtain

$$|||F_o(b)(Id-\Pi_2(\dot{I}_2))\ E(\hat{I})f|||^{\otimes} \leq c_9\ \beta_o^{3/5}\ |||f|||^{\otimes} \quad \forall f \in H^{\otimes}.$$

A similar argument for $F_o(b)(Id-\Pi_1(\dot{I}_1))\Pi_2(\dot{I}_2)E(\hat{I})$ leads to the result

$$|||F_o(b)(E(\hat{I})\ -\ E(\dot{I}))f|||^{\otimes} \leq c_{10}\ \beta_o^{3/5}\ |||f|||^{\otimes} \quad \forall f \in H^{\otimes}.$$

(3) Estimate of $\sum_{k=1}^{r} F_o(b)(E(b_k^+) + E(b_k^-))$:

It suffices to consider $F_o(b)E(b_k^+)$, $k \geq 2$, as the other terms can be handled similarly. Let ϑ_k^+, $\vartheta_k^* \in (-\frac{\pi}{2}, 0)$ be the angles between the lines joinint Λ_o, Λ_k^+ and Λ^*, Λ_k^+ and the λ_2-axis respectively. Then

$$|\vartheta_k^+ - \vartheta_k^*| \leq c_{11}\ \beta_o. \tag{10.21}$$

We decompose $F_o(b)E(b_k^+)$:

$$F_o(b)E(b_k^+) = F_o(b)\ \Delta_o^{-1}\ \Delta_o\ E(b_k^+)$$

$$= -F_o(b)\ \Delta_o^{-1}(V \circ \Theta(\vartheta_k^*))_{12}(V \circ \Theta(\vartheta_k^*))_{21}\ E(b_k^+) \tag{10.22}$$

$$+F_o(b)\ \Delta_o^{-1}(V \circ \Theta(\vartheta_k^+))_{11}(V \circ \Theta(\vartheta_k^+))_{22}\ E(b_k^+) + \Phi.$$

Using (10.21) the remainder term Φ can be estimated by

$$|||\Phi f|||^{\otimes} \leq c_{11}\ \beta_o\ |||E(b_k^+)f|||^{\otimes}.$$

We shall deal with the other two terms on the right-hand side of (10.22) separately.

(a) With $\vartheta = \vartheta_k^*$ we get from Corollary 7.2

$$(V \circ \Theta(\vartheta_k^*)) \circ \hat{\Theta}(\vartheta_k^*) \circ (\overline{\Gamma}-\Lambda^*) = T-V \circ \Lambda^* = (V \circ \Theta(\vartheta_k^*)) \circ \overline{\Gamma}(\vartheta_k^*,\Lambda^*).$$

The first component of this system yields

$$(V \circ \Theta(\vartheta_k^*))_{11}\ \overline{\Gamma}_1(\vartheta_k^*,\Lambda^*) + (V \circ \Theta(\vartheta_k^*))_{12}\ \overline{\Gamma}_2(\vartheta_k^*,\Lambda^*) = (T-V \circ \Lambda^*)_1$$

so that

$$(V \circ \Theta(\vartheta_o))_{11}^{-1}\ (V \circ \Theta(\vartheta_k^*))_{11}\ \overline{\Gamma}_1(\vartheta_k^*,\Lambda^*)$$

$$+\ (V \circ \Theta(\vartheta_o))_{11}^{-1}\ (V \circ \Theta(\vartheta_k^*))_{12}\ \overline{\Gamma}_2(\vartheta_k^*,\Lambda^*)$$

$$= T_{\vartheta_o,\Lambda^*,11} \quad \text{on} \quad \mathcal{R}(E(\hat{I})).$$

Since

$$T_{\vartheta_o,\Lambda^*,11}\ E_{\vartheta,\Lambda^*,11}\ (\{0\}) = 0$$

it follows that

$$T_{\vartheta_o,\Lambda^*,11}\ G_o(b) = 0$$

and we may conclude that

$$G_o(b)\ (V \circ \Theta(\vartheta_o))_{22}^{-1}\ (V \circ \Theta(\vartheta_o))_{11}^{-1}\ (V \circ \Theta(\vartheta_k^*))_{11}(V \circ \Theta(\vartheta_k^*))_{21}\overline{\Gamma}_1(\vartheta_k^*,\Lambda^*)$$

$$= -G_o(b)(V \circ \Theta(\vartheta_o))_{11}^{-1}\ (V \circ \Theta(\vartheta_o)_{22}^{-1}(V \circ \Theta(\vartheta_k^*))_{21}(V \circ \Theta(\vartheta_k^*))_{12}\overline{\Gamma}_2(\vartheta_k^*,\Lambda^*) \tag{10.23}$$

on $\mathcal{R}(E(\hat{I}))$. Now, using (10.23), the first term on the right-hand side of (10.22) can be seen to be equal to

$$(V(\vartheta_o)_{G_o(b)})^{-1} G_o(b)\ (V \circ \Theta(\vartheta_o))_{11}^{-1}\ (V \circ \Theta(\vartheta_o))_{22}^{-1} \cdot$$
$$\cdot (V \circ \Theta(\vartheta_k^*))_{11}\ (V \circ \Theta(\vartheta_k^*))_{21}\ \overline{\Gamma}_1(\vartheta_k^*,\Lambda^*)\ \overline{\Gamma}_2(\vartheta_k^*,\Lambda^*)^{-1}\ E(b_k^+). \tag{10.24}$$

The norm of this term may be estimated by

$$c\beta_o^{4/5}\ |||E(b_k^+)f|||^{\otimes},$$

where use has been made of (10.9):

$$|||\overline{\Gamma}_1(\vartheta_k^*,\Lambda^*)E(b_k^+)f||| \leq c\ k\ \beta_o^{8/5}\ |||E(b_k^+)f|||^{\otimes}$$
$$|||\overline{\Gamma}_2(\vartheta_k^*,\Lambda^*)^{-1}\ E(b_k^+)f||| \leq c(k\beta_o^{4/5})^{-1}\ |||E(b_k^+)f|||^{\otimes}. \tag{10.25}$$

(b) By setting $\vartheta = \vartheta_k^+$ and applying Corollary 7.2 we obtain a similar estimate for the second term in (10.22). Let

$$X = G_o(b)(V \circ \Theta(\vartheta_o))_{11}^{-1}\ (V \circ \Theta(\vartheta_o))_{22}^{-1}(V \circ \Theta(\vartheta_k^+))_{11}(V \circ \Theta(\vartheta_k^+))_{22}E(b_k^+).$$

The operator X satisfies the equation

$$X - T_{\vartheta_k^+,\Lambda_o,22}\ G_o(b)\ X\ \overline{\Gamma}_2(\vartheta_k^+,\Lambda_o)^{-1}\ E(b_k^+)$$
$$= T_{\vartheta_k^+,\Lambda_o,22}\ G_o(b)\ (V \circ \Theta(\vartheta_k^+))_{11}\ (V \circ \Theta(\vartheta_k^+))_{22}^{-1} \cdot$$
$$\cdot[((V \circ \Theta(\vartheta_o))_{22} - (V \circ \Theta(\vartheta_k^+))_{22}]\overline{\Gamma}_2(\vartheta_k^+,\Lambda_o)^{-1}\ E(b_k^+)$$
$$+ G_o(b)(V \circ \Theta(\vartheta_o))_{22}^{-1}\ (V \circ \Theta(\vartheta_k^+))_{21}\ \overline{\Gamma}_1(\vartheta_k^+,\Lambda_o)\ \overline{\Gamma}_2(\vartheta_k^+,\Lambda_o)^{-1}\ E(b_k^+).$$

Let $\|\cdot\|_{\vartheta,11,22}$ denote the norm in the space $H^{\otimes}_{\vartheta,11,22}$ defined by

$$H^{\otimes}_{\vartheta,11,22} := \{H^{\otimes},\ (\cdot,|(V \circ \Theta(\vartheta))_{11}(V \circ \Theta(\vartheta))_{22}|\cdot)\},\ -\frac{\pi}{2} < \vartheta < 0.$$

Then,

$$\|T_{\vartheta_k^+,\Lambda_o,22}\, G_o(b)f\|^{\otimes}_{\vartheta_o,11,22} \leq \beta_o \,\|f\|^{\otimes}_{\vartheta_o,11,22}, \quad f \in H^{\otimes}. \tag{10.27}$$

Moreover, using (10.8) we have

$$|||\bar{\Gamma}_2(\vartheta_k^+,\Lambda_o)^{-1}\, E(b_k^+)f|||^{\otimes} \leq c\,(k\beta_o^{4/5})^{-1}\, |||E(b_k^+)f|||^{\otimes}, \quad f \in H^{\otimes}. \tag{10.28}$$

From (10.27) and (10.28) we see that for $\hat{\beta}$ small enough the assumptions of Lemma 9.1 are satisfied. In particular, the first term Ω_1 on the right-hand side of (10.26) satisfies (note $|\vartheta_k^+ - \vartheta_o| \leq c\,k\,\beta_o^{4/5}$)

$$\|\Omega_1 f\|^{\otimes}_{\vartheta_o,11,22} \leq c\beta_o^{4/5}\,\|f\|^{\otimes}.$$

Thus, according to Lemma 9.1 and the estimate (7.10)

$$\|Xf\|^{\otimes}_{\vartheta_o,11,22} \leq c\beta_o^{4/5}\,\|E(b_k^+)f\|^{\otimes} \leq c\beta_o^{4/5}\,|||E(b_k^+)f|||^{\otimes} \qquad \forall f \in H^{\otimes}$$

Altogether we obtain

$$|||F_o(b)E(b_k^+)f|||^{\otimes} \leq c\beta_o^{4/5}\,|||E(b_k^+)f|||^{\otimes}, \qquad f \in H^{\otimes} \tag{10.29}$$

where the generic constant $c > 0$ can be chosen independently of k. Cauchy-Schwartz's inequality yields

$$\Big(|||\sum_{k=2}^{r} F_o(b)(E(b_k^+) + E(b_k^-))f|||^{\otimes}\Big)^2$$

$$= \sum_{k=2}^{r}\sum_{j=2}^{r}\,[(E(b_k^+) + E(b_k^-))f,\ F_o(b)(E(b_j^+) + E(b_j^-))f]^{\otimes}$$

$$\leq |||\sum_{k=2}^{r}(E(b_k^+) + E(b_k^-))f|||^{\otimes}\cdot|||\sum_{k=2}^{r} F_o(b)(E(b_k^+) + E(b_k^-))f|||^{\otimes}.$$

Now, by (10.29) and its analogue for b_k^-, we obtain

$$\left|\left|\left| \sum_{k=2}^{r} F_o(b)(E(b_k^+) + E(b_k^-))f \right|\right|\right|^{\otimes}$$

$$\leq c\beta_o^{4/5}\, r^{1/2} \left|\left|\left| \sum_{k=2}^{r} (E(b_k^+) + E(b_k^-))f \right|\right|\right|^{\otimes}$$

$$\leq c\beta_o^{3/5} \, |||f|||^{\otimes}$$

where the latter estimate follows since $\beta_o^{-2/5} \leq r < 1 + \beta_o^{-2/5}$.

(4) Estimate for $F_o(b)(E(b_1^+) + E(b_1^-))$:

The procedure is similar to the one used in (3).

We first decompose $F_o(b)E(b_1^+)$. If $\Lambda^* \neq \Lambda^+$, then $\vartheta_+^* \in (-\frac{\pi}{2}, 0)$ denotes the angle between the line connecting Λ^*, Λ^+ and the λ_2-axis. If $\Lambda^* = \Lambda^+$ then ϑ_+^* is chosen to be the angle between the tangent at Λ^+ and the λ_2-axis. With these conventions we get

$$\begin{aligned} F_o(b)E(b_1^+) &= F_o(b)\Delta_o^{-1}\, \Delta_o\, E(b_1^+) \\ &= -F_o(b)\, \Delta_o^{-1}(V \circ \Theta(\vartheta_+^*))_{12}(V \circ \Theta(\vartheta_+^*))_{21}\, E(b_1^+) \qquad (10.30) \\ &+ F_o(b)\, \Delta_o^{-1}\, (V \circ \Theta(\vartheta_o))_{11}\, (V \circ \Theta(\vartheta_o))_{22}\, E(b_1^+) + \Phi. \end{aligned}$$

Since $|\vartheta_+^* - \vartheta_o| < c'\, \beta_o$, the remainder term Φ satisfies

$$|||\Phi f|||^{\otimes} \leq c\beta_o\, |||f|||^{\otimes}$$

for some constant $c > 0$ and all $f \in H^{\otimes}$. Again we obtain the estimate for the first term on the right-hand side of (10.30) via an expression of the form (10.24) with $k = 1$. For the second term we consider

$$X = G_o(b)(Id+W_{o,b})\, E(b_1^+)$$

as a solution of the equation

$$X - T_{\vartheta_o,\Lambda_o,22}\, G_o(b)\, X\, \overline{\Gamma}_2(\vartheta_o,\Lambda_o)^{-1}\, E(b_1^+)$$

$$= G_o(b)(V\circ\Theta(\vartheta_o))_{21}(V\circ\Theta(\vartheta_o))_{22}^{-1}\, \overline{\Gamma}_1(\vartheta_o,\Lambda_o)(Id+W_{o,b})\overline{\Gamma}_2(\vartheta_o,\Lambda_o)^{-1}E(b_1^+) =: \Omega.$$

The assumptions of lemma 9.2 are satisfied:

$$|||\bar{\Gamma}_2(\vartheta_o,\Lambda_o)^{-1}\ E(b_1^+)f|||^{\otimes} \leq \beta_o^{-1}\ |||f|||^{\otimes},$$

$$\|T_{\vartheta_o,\Lambda_o,22}\,G_o(b)g\|^{\otimes}_{\vartheta_o,11,22} \leq \beta_o\ \|g\|^{\otimes}_{\vartheta_o,11,22}.$$

In particular

$$\|\Omega f\|^{\otimes}_{\vartheta_o,11,22} \leq c\,|||\bar{\Gamma}_1(\vartheta_o,\Lambda_o)(Id+W_{o,b})\bar{\Gamma}_2(\vartheta_o,\Lambda_o)^{-1}\ E(b_1^+)|||^{\otimes}$$

$$\leq c\beta_o^{3/5}\ |||E(b_1^+)f|||^{\otimes}.$$

With $W = (Id+W_{o,b})E(b_1^+)$ we may apply Lemma 9.2 and obtain as in (1)

$$\|Xf\|^{\otimes}_{\vartheta_o,11,22} \leq c(\varkappa)\ \beta_o^{3/5}\ |||(Id+W_{o,b})^{1+\varkappa}\ E(b_1^+)f|||^{\otimes}.$$

A similar result holds with b_- in place of b_+. Therefore, we obtain

$$|||F_o(b)(E(b_1^+) + E(b_1^-))f|||^{\otimes} \leq c(\varkappa)\ \beta_o^{3/5}\ (|||W^{\varkappa}_{o,b}f|||^{\otimes} + |||f|||^{\otimes}).$$

(5) Collecting the results of (1) - (4) and using the Cauchy-Schwartz inequality, we obtain the desired estimate in $H^{\otimes}$. Since all operators under consideration in H are direct sums of operators in $H^{\otimes}$, the result is obviously valid in H. □

The estimate of Lemma 10.2 plays a central role in the proof of the next theorem. In what follows we use the notation $(\cdot|\cdot)$ to denote points in $\mathbb{R}^2$ to distinguish these from open intervals in $\mathbb{R}$.

<u>Theorem 10.3:</u> For every finite left open, right closed arc b of an eigencurve C^1_ν whose closure does not contain intersection points with other eigencurves C^1_τ, $\nu \neq \tau$, there exists a sequence of inscribed polygons $\{P_n\}_{n=1}^{\infty}$ with vertices $\Lambda_{k,n} = (\lambda_{k,n}|\mu_{k,n})$, $k = 0,1,2,\dots,n+1$, $n \in \mathbb{N}$, on b with $\mu_{k,n} < \mu_{k+1,n}$, $k = 0,\dots,n$ and

$$\max_k |\Lambda_{k,n} - \Lambda_{k+1,n}| \to 0 \quad \text{as} \quad n \to \infty,$$

such that

$$E(b) = \lim_{n\to\infty} \sum_{k=0}^{n} F_o(b_{k,n}) \tag{10.31}$$

where $b_{k,n}$ denotes the arc of b connecting $\Lambda_{k,n}$ and $\Lambda_{k+1,n}$, $k = 0,\ldots,n$, $n \in \mathbb{N}$.

Proof: Let $\Lambda^{\pm} = (\Lambda^{\pm}|\mu^{\pm})$ denote the left and right endpoint of b. Then there is a slightly larger arc $\hat{b}$ still containing no intersection points with endpoints say $\hat{\Lambda}^{\pm} = (\hat{\Lambda}^{\pm}|\hat{\mu}^{\pm})$. Choose a rectangle $\hat{I}$ as in Lemma 10.2. Consider the interval

$$[\mu^-,\ \mu^+ + \frac{3}{2n}(\mu^+ - \mu^-)]$$

and define

$$\mu'_{k,n} = \mu^- + \frac{k}{2n}(\mu^+ - \mu^-),\quad k = 0,\ldots,2n+3,$$

for n sufficiently large. Apply Lemma 9.6 to

$$\rho(\mu) := \sum_{s=1}^{\infty} \frac{1}{s^2}(\varphi_s, \Pi_2(\mu)\varphi_s)$$

where $\{\varphi_s\}_{s=1}^{\infty}$ is a complete orthonormal system in $H^{\otimes}$.

For $(\lambda_1,\lambda_2) = (\mu'_{2k-2,n},\mu'_{2k+1,n})$ and $[\lambda'_1,\lambda'_2] = [\mu'_{2k-1,n},\mu'_{2k,n}]$ we deduce the existence of a point $\mu_{k,n} \in [\mu'_{2k-1,n}\ \mu'_{2k,n}]$ such that

$$\frac{\rho(\mu) - \rho(\mu_{k,n})}{\mu - \mu_{k,n}} \le \frac{4n}{\mu^+ - \mu^-}(\rho(\mu'_{2k+1,n}) - \rho(\mu'_{2k-2,n})) \tag{10.32}$$

for all $\mu \in [\mu'_{2k-2,n},\ \mu'_{2k+1,n}]$, $k = 1,\ldots,n+1$. Note that

$$[\tfrac{1}{2}(\mu_{k-1,n} + \mu_{k,n}),\ \tfrac{1}{2}(\mu_{k,n} + \mu_{k+1,n})] \subset [\mu'_{2k-2,n},\ \mu'_{2k+1,n}] \tag{10.33}$$

for $k = 2,\ldots,n-1$, and

$$\begin{aligned} &[\mu_{1,n},\ \tfrac{1}{2}(\mu_{1,n} + \mu_{2,n})] \subset [\mu^-,\mu'_{3,n}],\\ &[\tfrac{1}{2}(\mu_{n,n} + \mu_{n+1,n}),\ \mu_{n+1,n}] \subset [\mu'_{2n},\ \mu^+ + \tfrac{3}{2n}(\mu^+ - \mu^-)]. \end{aligned} \tag{10.34}$$

The points $\Lambda_{o,n} = \Lambda^-$, $\Lambda_{n+1,n} = \Lambda^+$ and $\Lambda_{k,n} = (\lambda_{k,n}|\mu_{k,n}) \in \hat{b}$, $k = 1,\ldots,n$, are the vertices of the polygon P_n. The $\lambda_{k,n}$ are uniquely determined by the $\mu_{k,n}$. If $b_{k,n}$ denotes the arc between $\Lambda_{k,n}$ and $\Lambda_{k+1,n}$, $k = 1,\ldots,n-1$ and $b_{n,n}$ the arc between $\Lambda_{n,n}$ and $(\lambda_{n+1,n}|\mu_{n+1,n})$ then we have, according to Lemma 10.2 with $\varkappa = \frac{1}{4}$,

$$
\begin{aligned}
&|[g, \sum_{k=1}^{n} (E(b_{k,n}) - F_o(b_{k,n}))\, E(\hat{I})f]^{\otimes}| \\
&\leq cn^{-3/5} \{|||g|||^{\otimes} \sum_{k=1}^{n} |||(W_{o,b_{k,n}})^{1/4}\, E(b_{k,n})f|||^{\otimes} \\
&\quad + \sum_{k=1}^{n} |||F_o(b_{k,n})g|||^{\otimes} (|||f|||^{\otimes} + |||(W_{o,b_{k,n}})^{1/4}\, f|||^{\otimes})\} \qquad (10.35) \\
&\leq cn^{-1/10} \{|||g|||^{\otimes} (\sum_{k=1}^{n} (|||(W_{o,b_{k,n}})^{1/4}\, E(b_{k,n})f|||^{\otimes})^2)^{1/2} \\
&\quad + |||f|||^{\otimes} (\sum_{k=1}^{n} (|||F_o(b_{k,n})g|||^{\otimes})^2)^{1/2}\}.
\end{aligned}
$$

Now, using Lemma 9.8 we can show that

$$\sum_{k=1}^{n} F_o(b_{k,n}) \text{ is bounded uniformly with respect to } n. \qquad (10.36)$$

To see this, first observe that

$$E_{\vartheta_o,\Lambda^*,11}(\{0\})\, F_o(b_{k,n}) = E_{\vartheta_o,\Lambda_o,22}((-\beta_o,\beta_o])\, F_o(b_{k,n}) = F_o(b_{k,n})$$

by the construction of $F_o(b_{k,n})$. Since V_{11}, V_{12} are invertible on the range of $G_o(b_{k,n})$ by the compactness assumption on T_1 and since $V_{11}\, V_{12} \leq \Delta_o$, it follows that Δ_o is bounded invertible on $R(G_o(b_{k,n})) = R(F_o(b_{k,n}))$ and we get the estimates

$$
\begin{aligned}
&|||\Delta_o^{-1/2}(T-V \circ \Lambda_{k,n})_i\, F_o(b_{k,n})v|||^{\otimes} = \|(T-V \circ \Lambda_{k,n})_i\, F_o(b_{k,n})v\|^{\otimes} \\
&\leq c|\mu_{k+1,n} - \mu_{k,n}|\; |||F_o(b_{k,n})v|||^{\otimes}, \quad i = 1,2,
\end{aligned}
$$

and

$$|||\Delta_o^{-1/2} F_o(b_{k,n})v||| \leq |||F_o(b_{k,n})v|||^{\otimes}$$

where the generic constant c depends only on $\hat{b}$ and not on k,n. Thus the following calculation is justified:

$$(\mu_{k,n} - \mu_{1,n}) [F_o(b_{k,n})f, F_o(b_{1,n})g]^{\otimes}$$

$$= -((\Delta_2 - \Delta_o \mu_{k,n}) F_o(b_{k,n})f, F_o(b_{1,n})g)^{\otimes}$$

$$+ (F_o(b_{k,n})f, (\Delta_2 - \Delta_o \mu_{1,n}) F_o(b_{1,n})g)^{\otimes}$$

$$= -[\Delta_o^{-1/2}(\Delta_2 - \Delta_o \mu_{k,n}) F_o(b_{k,n})f, F_o(b_{1,n})g]^{\otimes}$$

$$+ [F_o(b_{k,n})f, \Delta^{-1/2}(\Delta_2 - \Delta_o \mu_{1,n}) F_o(b_{1,n})g]^{\otimes}.$$

Consequently, with

$$Z_k = F_o(b_{k,n})$$
$$M_k = -\Delta_o^{-1/2}(\Delta_2 - \Delta_o \mu_{k,n}) F_o(b_{k,n})$$
$$N_k = \Delta_o^{-1/2} F_o(b_{k,n})$$
$$\gamma_k = \mu_{k,n}, \quad k = 1,\dots,n,$$

the assumptions of Lemma 9.8 are satisfied, and (10.36) follows. With this result we get from (10.35) that

$$|[g, \sum_{k=1}^{n} (E(b_{k,n}) - F_o(b_{k,n})) E(\hat{I})f]^{\otimes}|$$
$$\leq cn^{-1/10} \{(\sum_{k=1}^{n} (|||(W_{o,b_{k,n}})^{1/4} E(b_{k,n})f|||^{\otimes})^2)^{1/2} + |||f|||^{\otimes}\} |||g|||^{\otimes} \qquad (10.37)$$

for $f,g \in H^{\otimes}$.

It remains to estimate the sum term in (10.37). Applying the spectral theorem we may write

$$|[g, \sum_{k=1}^{n} (E(b_{k,n}) - F_o(b_{k,n}) E(\hat{I})f]^{\otimes}| \tag{10.38}$$

$$\leq cn^{-1/10} \{ (\int_{\mu_{1,n}}^{\mu_{n+1,n}} \eta_n(\mu) \, d[f, \Pi_2(\mu)f]^{\otimes})^{1/2} + |||f|||^{\otimes}\} \, |||g|||^{\otimes}$$

where η_n is defined by

$$\eta_n(\mu) = [\sin \pi \frac{\mu - \mu_{k,n}}{\mu_{k+1,n} - \mu_{k,n}}]^{-1/2} \quad \text{for } \mu \in [\mu_{k,n}, \mu_{k+1,n}].$$

Now, using (10.32) - (10.34) and integration by parts, we obtain

$$\int_{\lambda_{1,n}}^{\lambda_{n+1,n}} \eta_n(\mu) \, d\rho(\mu) \leq c(\rho(\mu^+ + \frac{3}{2n} (\mu^+ - \mu^-)) - \rho(\mu^-))$$

where c is independent of n. The boundedness of the term in question follows for $f \in H^{\otimes}$ such that $\frac{d[f, \Pi_2(\mu)f]^{\otimes}}{d\rho(\mu)}$ is bounded.

According to Lemma 9.5 the set of all such f, denoted by $H^{\otimes}_{\rho}$, is dense in $H^{\otimes}$. Since g is arbitrary we have for $f \in H^{\otimes}_{\rho}$

$$||| \sum_{k=1}^{n} (E(b_{k,n}) - F_o(b_{k,n})) E(\hat{I})f |||^{\otimes} \leq c(f) \, n^{-1/10}. \tag{10.39}$$

Now, let $\{\hat{I}_k\}_{k=1}^{\infty}$ be a monotone sequence of rectangle of the same type as $\hat{I}$ with

$$\hat{I}_k \to \mathbb{R}^2 \quad \text{as} \quad k \to \infty.$$

Moreover, let the orthonormal system used to define ρ be such that for every $j \in \mathbb{N}$

$$\varphi_j \in R(E(\hat{I}_k)) \text{ for k sufficiently large.}$$

Then it follows that (10.39) remains true for any fixed $f \in H^{\otimes}$ which is in the span of the $\{\varphi_j\}_{j=1}^{\infty}$. Since

$$\operatorname*{s-lim}_{n\to\infty} \sum_{i=1}^{n} E(b_{k,n}) = E(b)$$

it follows that $\lim_{n\to\infty} \sum_{k=1}^{n} F_o(b_{k,n})f$ exists and

$$E(b)f = \lim_{n\to\infty} \sum_{k=1}^{n} F_o(b_{k,n})f \quad \forall f \in Sp[\varphi_j, j = 1,2,3,\ldots] \tag{10.40}$$

Since $Sp[\varphi_j,\ j = 1,2,3,\ldots]$ is dense in $H^{\otimes}$ it follows from (10.36) that (10.40) remains valid for any $f \in H^{\otimes}$. Let $b'_{k,n} = b_{k,n}$, $k = 1,\ldots,n-1$, $b'_{o,n}$ denote the arc between Λ^- and $\Lambda_{1,n}$ and let $b'_{n,n}$ denote the arc between $\Lambda_{n,n}$ and Λ^+. Then

$$\sum_{k=0}^{n} F_o(b'_{k,n}) = \sum_{k=1}^{n} F_o(b_{k,n}) + F_o(b'_{o,n}) - F_o(b_{n,n}) + F_o(b'_{n,n})$$

Application of Lemma 9.7 to the second factor in the definition of $G_o(b_{n,n})$ and from Rellich's theorem it follows that

$$\operatorname*{s-lim}_{n\to\infty} G_o(b_{n,n}) = E_{\vartheta_o,\Lambda^+,11}(\{0\})\ E_{\vartheta_o,\Lambda^+,22}(\{0\}) \tag{10.41}$$

where $\vartheta_o \in (-\frac{\pi}{2}, 0)$ is the tangent angle at Λ^+. Since as a consequence of the multiparameter theorem (Theorem 7.1) common eigensolutions of $\overline{\Gamma}_1$, $\overline{\Gamma}_2$ for an eigenvalue Λ are in the null space of $T - V \circ \Lambda$ and vice versa, we have

$$E_{\vartheta_o,\Lambda^+,11}(\{0\})\ E_{\vartheta_o,\Lambda^+,22}(\{0\})\ H^{\otimes}_{\vartheta_o,11,22} = E(\{\Lambda^+\})H^{\otimes}.$$

Thus

$$\operatorname*{s-lim}_{n\to\infty} F_o(b_{n,n}) = E(\{\Lambda^+\}). \tag{10.42}$$

Similarly,

$$\operatorname*{s-lim}_{n\to\infty} F_o(b'_{n,n}) = E(\{\Lambda^+\}). \tag{10.43}$$

On the other hand by the same argument

$$\operatorname*{s-lim}_{n\to\infty} F_o(b'_{o,n}) = 0.$$

This finally proves the theorem. □

Because of the integral type character of the limit, following Cordes we write

$$\int_b F_0(ds) := \lim_{n\to\infty} \sum_{k=0}^{n} F_0(b_{k,n}) \tag{10.44}$$

for the limit occurring in the theorem. It should be kept in mind, however, that the theorem gives basically an existence theorem for a suitable polygon approximation. The choice of corner points depends on the spectral properties of $\bar{\Gamma}_2$ and is usually not arbitrary.

If Λ^0 is an intersection point of C^1_ν with some other eigencurve and b^0 is an arc on C^1_ν containing Λ^0, we modify the definition of $G_0(b^0)$ by taking the orthogonal projection on the closure of

$$Sp[S_\nu(\Lambda^0)] \subset H^{\otimes}_{\vartheta_0,11}$$

denoted by

$$E^\nu_{\vartheta_0,\Lambda^0,11}(\{0\})$$

rather than the projection

$$E_{\vartheta_0,\Lambda^0,11}(\{0\})$$

on the full null space $N(T_{\vartheta_0,\Lambda^0,11})$.

This way the result of the last theorem may be extended to include intersection points. Let the corresponding modified projection again be called $F_0(\cdot)$. Since, at a point of intersection Λ^0, the sum of projection

$$\sum_{\nu:\Lambda^0 \in C'_\nu} E^\nu_{\vartheta_0,\Lambda^0,11}(\{0\})$$

maps onto the whole nullspace we have:

<u>Corollary 10.4:</u> Let b be a left closed, right open arc on an eigencurve C^1_ν then

$$E(b) = \int_b F_0(ds).$$

Extending the notation to the direct sum $H = H^{\otimes} \oplus H^{\otimes}$ finally yields

Corollary 10.5: For any rectangle $I \subset \mathbb{R}^2$ we have

$$\Pi(I) = \sum_{\nu} \int_{I \cap C_{\nu}^{1}} F(ds).$$

Proof: Noting that for finite rectangles I the sum is finite, we immediately see that the result is true as a consequence of Corollary 10.4. For unbounded rectangles the integral has to be understood, as customary in spectral theory, in the sense of strong limits of integrals over finite regions. Then the result follows again from the equality in Corollary 10.4.

□

The result gives a spectral representation for the two parameter system in terms of spectral properties of the original operators since the projections $F(\cdot)$ are defined in terms of the spectral measures of weighted versions of T_1 and T_2. To our knowledge, this has not been achieved for any other multiparameter spectral representation theorem.

References

1. F.V. Atkinson, Multiparameter spectral theory, Bull. Amer. Math. Soc. 74, 1-28, 1968.
2. F.V. Atkinson, Multiparameter Eigenvalue Problems, Vol.I: Matrices and Compact Operators, Academic Press, New York, 1972.
3. N.I. Akhiezer and I.M. Glazman, Theory of Linear Operators in Hilbert Space, Vol.1, Pitman, London, 1981.
4. Yu. M. Berezanskii, Expansions in Eigenfunctions of Self-Adjoint Operators, Translations of Mathematical Monographs 17, Amer. Math. Soc Providence, Rhode Island, 1968.
5. H.O. Cordes, Separation der Variablen in Hilbertschen Räumen, Math. Ann. 125, 401-434, 1953.
6. H.O. Cordes, Über die Spektralzerlegung von hypermaximalen Operatoren die durch Separation der Variablen zerfallen, I, Math. Ann. 128, 257-289, 1954; II, Math. Ann. 128, 373-411, 1955.
7. F. Rellich, Störungstheorie der Spektralzerlegung, V. Mitteilung, Math Ann. 118, 677-685, 1937.
8. J. Schur, Bemerkungen zur Theorie der beschränkten Bilinearformen mit unendlich vielen Veränderlichen, J. Reine Angew. Math. 140, 1-28, 1911.
9. B.D. Sleeman, Multiparameter Spectral Theory in Hilbert Space, Pitman, London, 1978.
10. H. Volkmer, On Multiparameter theory, J. Math. Anal. Appl. 86, 44-53, 1982.